GUIDELINES FOR
CHEMICAL REACTIVITY
EVALUATION
AND APPLICATION TO
PROCESS DESIGN

Publications Available from the
CENTER FOR CHEMICAL PROCESS SAFETY
of the
AMERICAN INSTITUTE OF CHEMICAL ENGINEERS

Guidelines for Process Safety Fundamantals in General Plant Operations
Guidelines for Chemical Reactivity Evaluation and Application to Process Design
Tools for Making Acute Risk Decisions with Chemical Process Safety Applications
Guidelines for Preventing Human Error in Process Safety
Guidelines for Evaluating the Characteristics of Vapor Cloud Explosions, Flash Fires, and BLEVEs
Guidelines for Implementing Process Safety Management Systems
Guidelines for Safe Automation of Chemical Processes
Guidelines for Engineering Design for Process Safety
Guidelines for Auditing Process Safety Management Systems
Guidelines for Investigating Chemical Process Incidents
Guidelines for Hazard Evaluation Procedures, Second Edition with Worked Examples
Plant Guidelines for Technical Management of Chemical Process Safety, Rev. Ed.
Guidelines for Technical Management of Chemical Process Safety
Guidelines for Chemical Process Quantitative Risk Analysis
Guidelines for Process Equipment Reliability Data, with Data Tables
Guidelines for Vapor Release Mitigation
Guidelines for Safe Storage and Handling of High Toxic Hazard Materials
Guidelines for Use of Vapor Cloud Dispersion Models
Safety, Health, and Loss Prevention in Chemical Processes: Problems for Undergraduate Engineering Curricula
Safety, Health, and Loss Prevention in Chemical Processes: Problems for Undergraduate Engineering Curricula—Instructor's Guide
Workbook of Test Cases for Vapor Cloud Source Dispersion Models
Proceedings of the International Symposium and Workshop on Safe Chemical Process Automation, 1994
Proceedings of the International Process Safety Management Conference and Workshop, 1993
Proceedings of the International Conference on Hazard Identification and Risk Analysis, Human Factors, and Human Reliability in Process Safety, 1992
Proceedings of the International Conference/Workshop on Modeling and Mitigating the Consequences of Accidental Releases of Hazardous Materials, 1991.
Proceedings of the International Symposium on Runaway Reactions, 1989
CCPS/AIChE Directory of Chemical Process Safety Services

GUIDELINES FOR CHEMICAL REACTIVITY EVALUATION AND APPLICATION TO PROCESS DESIGN

CENTER FOR CHEMICAL PROCESS SAFETY
of the
AMERICAN INSTITUTE OF CHEMICAL ENGINEERS
345 East 47th Street • New York, NY 10017

Copyright © 1995
American Institute of Chemical Engineers
345 East 47th Street
New York, New York 10017

All rights reserved. No part of this publication may be reproduced, stored in a retrieval system, or transmitted in any form or by any means, electronic, mechanical, photocopying, recording, or otherwise without the prior permission of the copyright owner.

Library of Congress Cataloging-in Publication Data
Guidelines for chemical reactivity evaluation and application to
　process design.
　　　p.　　cm.
　Includes bibliographic references and index.
　ISBN 0-8169-0479-0
　1. Chemical processes.　2. Reactivity (Chemistry).　I. American
Institute of Chemical Engineers. Center for Chemical Process
Safety.
TP155.7.G84 1995
680'. 2812—dc20　　　　　　　　　　　　　　　　　92–38794
　　　　　　　　　　　　　　　　　　　　　　　　　　CIP

This book is available at a special discount when ordered in bulk quantities. For information, contact the Center for Chemical Process Safety of the American Institute of Chemical Engineers at the address shown above.

It is sincerely hoped that the information presented in this document will lead to an even more impressive safety record for the entire industry; however, the American Institute of Chemical Engineers, its consultants, CCPS subcommittee members, their employers, their employers' officers and directors, and TNO Prins Maurits Laboratory disclaim making or giving any warranties or representations, express or implied, including with respect to fitness, intended purpose, use or merchantability and/or correctness or accuracy of the content of the information presented in this document. As between (1) the American Institute of Chemical Engineers, its consultants, CCPS subcommittee members, their employers, their employers' officers and directors, and TNO Prins Maurits Laboratory and (2) the user of this document, the user accepts any legal liability or responsibility whatsoever for the consequence of its use or misuse.

CONTENTS

List of Tables ix
List of Figures xi
Preface xv
Acknowledgments xvii
Glossary xix
List of Symbols xxv

1. INTRODUCTION 1

1.1 GENERAL 1
1.2 CHEMICAL REACTIVITY 4
1.3 DETONATIONS, DEFLAGRATIONS, AND RUNAWAYS 5
1.4 ASSESSMENT AND TESTING STRATEGIES 6

2. IDENTIFICATION OF HAZARDOUS CHEMICAL REACTIVITY 9

2.1. SUMMARY/STRATEGY 9
   2.1.1 Introduction 9
   2.1.2 Hazard Identification Strategy 9
   2.1.3 Exothermic Reactions 11
   2.1.4 Experimental Thermal and Reactivity Measurements 13
   2.1.5 Test Strategies 13
   2.1.6 Overview of Thermal Stability Test Methods 20
   2.1.7 Examples of Interpretation and Application of Test Data 22
2.2 TECHNICAL SECTION 28

2.2.2 Identification of High Energy Substances	30
2.2.3. Hazard Prediction by Thermodynamic Calculations	33
2.2.3.1 Oxygen Balance	*33*
2.2.3.2 Calculation of the Reaction Enthalpy	*35*
2.2.3.3 Application of Computer Programs	*39*
2.2.4 Instability/Incompatibility Factors	46
2.2.4.1 Factors Influencing Stability	*46*
2.2.4.2 Redox Systems	*49*
2.2.4.3 Reactions with Water	*51*
2.2.4.4 Reactions between Halogenated Hydrocarbons and Metals	*52*
2.3 PRACTICAL TESTING	**52**
2.3.1 Screening Tests	52
2.3.1.1 Thermal Analysis	*52*
2.3.1.2 Isoperibolic Calorimetry	*59*
2.3.2 Thermal Stability and Runaway Testing	61
2.3.2.1 Isothermal Storage Tests	*62*
2.3.2.2 Dewar Flask Testing and Adiabatic Storage Tests	*66*
2.3.2.3 Accelerating Rate Calorimeter (ARC)	*71*
2.3.2.4 Stability Tests for Powders	*76*
2.3.3 Explosibility Testing	78
2.3.3.1 Detonation Testing	*78*
2.3.3.2 Deflagration Testing and Autoclave Testing	*80*
2.3.3.3 Mechanical Sensitivity Testing	*83*
2.3.3.4 Sensitivity to Heating under Confinement	*86*
2.3.4 Reactivity Testing	87
2.3.4.1 Pyrophoric Properties	*87*
2.3.4.2 Reactivity with Water	*87*
2.3.4.3 Oxidizing Properties	*87*
2.3.5 Flammability Testing	88
3. CHEMICAL REACTIVITY CONSIDERATIONS IN PROCESS/REACTOR DESIGN AND OPERATION	**89**
3.1 INTRODUCTION	**89**
3.1.1 Thermal Hazards: Identification and Analysis	90
3.1.1.1 Cause, Definition, and Prevention of a Runaway	*90*
3.1.1.2 Some Simple Rules for Inherent Safety	*96*
3.1.1.3 Strategy for Inherent Safety in Design and Operation	*97*
3.1.1.4 Equipment to be Used for the Analysis of Hazards	*100*
3.2 REACTOR, HEAT AND MASS BALANCE CONSIDERATIONS	**100**

3.2.1 Heat and Mass Balances, Kinetics, and Reaction Stability	**100**
3.2.1.1 Adiabatic Temperature Rise	*101*
3.2.1.2 The Reaction	*102*
3.2.1.3 Reaction Rate	*102*
3.2.1.4 Reaction Rate Constant	*103*
3.2.1.5 Concentration of Reactants	*104*
3.2.1.6 Effect of Surrounding Temperature on Stability	*104*
3.2.1.7 Effect of Agitation and Surface Fouling on Stability	*106*
3.2.1.8 Mass Balance	*107*
3.2.2 Choice of Reactor	**108**
3.2.3 Heat Transfer	**113**
3.2.3.1 Heat Transfer in Nonagitated Vessels	*114*
3.2.3.2 Heat Transfer in Agitated Vessels	*114*
3.3 ACQUISITION AND USE OF PROCESS DESIGN DATA	**116**
3.3.1 Introduction	**116**
3.3.2 Bench-Scale Equipment for Batch/Tank Reactors	**116**
3.3.2.1 Reaction Calorimeter (RC1)	*117*
3.3.2.2 Contalab	*119*
3.3.2.3 CPA ThermoMetric Instruments	*121*
3.3.2.4 Quantitative Reaction Calorimeter	*122*
3.3.2.5 Specialized Reactors	*123*
3.3.2.6 Vent Size Package (VSP)	*124*
3.3.2.7 Reactive System Screening Tool (RSST)	*126*
3.3.3 Process Safety for Reactive Systems	**129**
3.3.3.1 Test Plan	*129*
3.3.3.2 System under Investigation	*131*
3.3.3.3 Test Results	*132*
3.3.3.4 Malfunction and Process Deviation Testing	*134*
3.3.3.5 Pressure Effect	*137*
3.3.3.6 Results from the ARC, RSST, and VSP	*137*
3.3.4 Scale-up and Pilot Plants	**137**
3.3.4.1 General Remarks	*137*
3.3.4.2 Chemical Kinetics	*139*
3.3.4.3 Mass Transfer/Mixing	*140*
3.3.4.4 Heat Transfer	*141*
3.3.4.5 Self-Heating	*142*
3.3.4.6 Scale-up of Accelerating Rate Calorimeter (ARC) Results	*145*
3.3.4.7 Scale-up of Vent Size Package (VSP) Results	*145*
3.3.5 Process Design Applications	**147**
3.3.5.1 Batch and Semi-Batch Processing Plants	*148*
3.3.5.2 An Example Involving Peroxides	*149*
3.3.5.3 An Example Involving a Continuous Nitration	*151*

3.3.5.4 A Self-Heating Example	*153*
3.3.5.5 Batch-to-Continuous Example	*154*
3.3.5.6 Integrated Relief Evaluation	*154*
3.3.6 Storage and Handling	**154**
3.3.6.1 Scale-up Example for Storage	*154*
3.3.6.2 Peroxides	*155*
3.3.6.3 Passive Means to Prevent Explosions	*156*
3.3.7 Dryers and Filters	**157**
3.4 PROTECTIVE MEASURES	**159**
3.4.1 Containment	**159**
3.4.1.1 Introduction	*159*
3.4.1.2 Determination of Gas–Vapor Release	*160*
3.4.1.3 Laboratory Scale	*161*
3.4.1.4 Full-Scale Example	*164*
3.4.2 Instrumentation and Detection of Runaways	**164**
3.4.2.1 Methods of On-Line Detection	*164*
3.4.2.2 Methods of Noise Suppression	*167*
3.4.3 Mitigation Measures	**168**
3.4.3.1 Reaction Quenching Methods	*168*
3.4.3.2 An Example Involving a Sulfonation	*169*
3.4.3.3 Relief Disposal	*170*
3.4.3.4 Dispersion, Flaring, Scrubbing, and Containment	*172*
3.4.3.5 Venting	*173*
4. MANAGEMENT OF CHEMICAL PROCESS SAFETY	**175**
4.1 HAZARD IDENTIFICATION AND QUANTIFICATION	**175**
4.2 HAZARD EVALUATION PROCEDURES	**176**
4.3 CHEMICAL PROCESS SAFETY MANAGEMENT	**180**
4.4 FUTURE TRENDS	**181**
REFERENCES	**183**
REFERENCES CITED	**183**
SELECTED ADDITIONAL READINGS	**198**
INDEX	**201**

LIST OF TABLES

TABLE 1.1	Suggested Stages in Assessment of Reactivity by Scale	6
TABLE 1.2	Typical Testing Procedures by Chronology	7
TABLE 2.1	Overview and Comparison of Calorimetric Techniques	21
TABLE 2.2	Comparison of T_o and ΔH_d for TBPB Using Different Calorimetric Techniques	24
TABLE 2.3	Example of Stability/Runaway Hazard Assessment Data and Evaluation Report	27
TABLE 2.4	Structure of High Energy Release Compounds	30
TABLE 2.5	Typical High Energy Molecular Structures	32
TABLE 2.6	Some Available Sources of Enthalpy of Formation Data	36
TABLE 2.7	Enthalpies of Formation (in kcal/mol) of 10 Chemicals Calculated by Five Methods at Standard Conditions of 20°C and 1 Bar	37
TABLE 2.8	Decomposition Products of t-Butylperoxybenzoate (TBPB)	38
TABLE 2.9	Comparison of Four Thermodynamic Calculation Computer Programs	41
TABLE 2.10	Enthalpy of Decomposition or Reaction	42
TABLE 2.11	Degree of Hazard in Relation to the Oxygen Balance (CHETAH Criterion 3)	44
TABLE 2.12	Degree of Hazard in Relation to the Y-Factor (CHETAH Criterion 4)	44
TABLE 2.13	Advantages and Disadvantages of REITP2	45
TABLE 2.14	Examples of Hazardous Incompatibility Combinations	47
TABLE 2.15	Structures Susceptible to Peroxidation in Presence of Air	50
TABLE 3.1	Vapor Pressure of Acetone at Different Temperatures	108
TABLE 3.2	Comparison of Different Reactor Types from the Safety Perspective	110

TABLE 3.3	Characteristics of the RSST and VSP	129
TABLE 3.4	Essential Questions on Safety Aspects of Reactions	130
TABLE 3.5	Reactor Scale-up Characteristics	140
TABLE 3.6	Combinations of Parameter Sensitivities	163

LIST OF FIGURES

FIGURE 1.1.	Key Parameters That Determine Design of Safe Chemical Plants	3
FIGURE 2.1.	Initial Theoretical Hazard Identification Strategy	10
FIGURE 2.2.	Types of Explosions	12
FIGURE 2.3.	Flow Chart for Preliminary Hazard Evaluation	14
FIGURE 2.4.	Flow Chart for a Strategy for Stability Testing	18
FIGURE 2.5.	Flow Chart for Specific Experimental Hazard Evaluation for Reactive Substances	20
FIGURE 2.6.	Typical Curves Obtained from: (A) Constant Heating Rate Tests, (B) Isothermal Tests, (C) Differential Thermal Analysis and (D) Adiabatic Calorimetry	23
FIGURE 2.7.	Depletion of Inhibitor Stability: DSC Curve (A) and Isothermal Curves (B) for an Inhibited Material.	25
FIGURE 2.8.	Typical Results of Autocatalytic Thermokinetics as Obtained by Isothermal Analysis.	26
FIGURE 2.9.	Schematic Energy Diagram of the Transition State Leading to Chemical Reaction	29
FIGURE 2.10.	Combination of Criteria 1 and 2 for Evaluating Explosibility in the CHETAH Program	43
FIGURE 2.11.	Reaction Rate as a Function of Temperature (Arrhenius Equation)	48
FIGURE 2.12.	Schematic Representation of Heat-flux DTA and Power Compensation DSC	53
FIGURE 2.13.	Example Scanning DSC Curve of an Exothermic Decomposition	55
FIGURE 2.14.	DSC Curve—Typical Exothermic Reaction	57

FIGURE 2.15.	DSC Curve—Steep Exothermic Rise	58
FIGURE 2.16.	Typical Isoperibolic Measurement	60
FIGURE 2.17.	Cross-Section of an Isothermal Storage Test (IST).	62
FIGURE 2.18.	Rate of Heat Generation (q) of Three Isothermal Experiments as a Function of Time (t) at Three Temperatures (T)	65
FIGURE 2.19.	Rate of Heat Generation as a Function of Temperature at Points of Isoconversion as Derived from Figure 2.18	66
FIGURE 2.20.	Simple Test Setup for a Dewar Flask Test	67
FIGURE 2.21.	Typical Temperature–Time Curves of Dewar Vessel Tests	68
FIGURE 2.22.	Arrangement of the Adiabatic Storage Test (AST)	69
FIGURE 2.23.	Adiabatic Induction Time	70
FIGURE 2.24.	Accelerating Rate Calorimeter (ARC)	72
FIGURE 2.25.	The Heat–Wait–Search Operation Mode of the ARC	73
FIGURE 2.26.	ARC Plot of Self-Heat Rate as a Function of Temperature.	74
FIGURE 2.27.	Heat Release Rate and Heat Transfer Rate versus Temperature	75
FIGURE 2.28.	Test Set-up of the TNO 50/70 Steel Tube Test	79
FIGURE 2.29.	Deflagration Rate of TBPB at Different Temperatures as a Function of Pressure Established in the CPA	81
FIGURE 2.30.	The BAM Friction Apparatus: Horizontal and Vertical Cross-Sections	84
FIGURE 2.31.	Bureau of Mines Impact Apparatus.	85
FIGURE 2.32.	Set-up of the Koenen Test	86
FIGURE 3.1.	Typical Heat Generation and Heat Removal Rates as a Function of Temperature	92
FIGURE 3.2.	Relation between Critical Heat Production Rates of Small Scale and of Plant Scale	95
FIGURE 3.3.	Comparison of Critical Temperatures for Frank-Kamenetskii and Semenov Models (Right Cylinder Configuration)	95
FIGURE 3.4.	Process Hazard Evaluation Scheme	98
FIGURE 3.5.	Methods to Reduce the Heat Production q	102
FIGURE 3.6.	Reaction Rate Constant k of a Reaction as a Function of Temperature	103
FIGURE 3.7.	Stability as a Function of Heat Production and Heat Removal.	105
FIGURE 3.8.	Effect of Agitation and Surface Fouling on Heat Transfer and Stability	107
FIGURE 3.9.	Example Reaction: Selectivity versus Temperature	111

LIST OF FIGURES

FIGURE 3.10.	Effect of T_0 in a Semi-Batch Reactor	113
FIGURE 3.11.	Modular Design of the Bench-Scale Reactor (RC1)	120
FIGURE 3.12.	Schematic Design of the Contalab	121
FIGURE 3.13.	The CPA System	122
FIGURE 3.14.	Sketch of the Quantitative Reaction Calorimeter	123
FIGURE 3.15.	Vent Size Package (VSP) Test Cell	125
FIGURE 3.16.	Schematic of the RSST Showing the Glass Test Cell and the Containment Vessel	127
FIGURE 3.17.	Typical Temperature–Time Curve of an RSST Experiment	128
FIGURE 3.18.	Semi-Batch versus Batch Operations for First- and Second-Order Kinetics	133
FIGURE 3.19.	Typical Structure for Reactor Design	139
FIGURE 3.20.	Typical Temperature Distributions during Self-Heating in a Vessel	143
FIGURE 3.21.	An Approach to Emergency Relief System Sizing in Case Necessary Kinetic and Thermophysical Data Are Lacking	147
FIGURE 3.22.	One Design for Safe Atmospheric Storage of Flammable Liquids	158
FIGURE 3.23.	Calculated and Measured Temperatures in a Layer as a Result of the Self-Heating of Tapioca.	158
FIGURE 3.24.	Flow Sheet to Determine Proper Site for Reactivity Testing (Laboratory or High-pressure Cell)	162
FIGURE 3.25.	Concept of Restabilization and Venting	167
FIGURE 3.26.	Decision Tree for Relief Disposal	171
FIGURE 4.1.	Example of a Fault Tree	178
FIGURE 4.2.	*F-n* Curve (Risk Curve)	179

PREFACE

The American Institute of Chemical Engineers has a long history of involvement with process safety and loss prevention in the chemical, petrochemical, petroleum, and other process industries. Through its strong link with process engineers, process designers, operating engineers, safety professionals, research and development engineers, managers, and academia,, the AIChE has enhanced communications and fostered improvements in the high safety standards established in the process industries. Publications, symposia,, and continuing education courses of the Institute are information resources for the engineering profession and for managers on the causes of industrial accidents and the means of prevention.

Early in 1985, the AIChE established the Center for Chemical Process Safety (CCPS) as a scientific and educational organization to provide expert leadership and focus on engineering practices and research that can prevent or mitigate catastrophic events involving hazardous materials. The first program to meet this objective was the initiation of the development of a series of Guidelines books covering a wide range of engineering practices and management techniques. The selection of the appropriate topics for Guidelines books is one role of the CCPS Technical Steering Committee, which consists of selected experts from sponsor organizations. The Technical Steering Committee considered a Guidelines document covering reactive chemicals as an essential element for this series of books.

A Reactive Chemicals Subcommittee was formed with the following members:

George T. Wildman, Chair, *Merck Chemical Manufacturing Division*
Glenn T. Bodman, *Eastman Kodak Company*
Louis P. Bosanquet, *Monsanto Chemical Company*
Donald J. Connolley, *Akzo Chemicals, Inc.*
Edward Donoghue, *American Cyanamid*
David V. Eberhardt, *Rohm and Haas Company*
James G. Hansel, *Air Products & Chemicals Company*

Horace E. Hood, *Hercules, Inc.*
Thomas Hoppe, *Ciba-Geigy Corporation*
Henry T. Kohlbrand, *Dow Chemical Company*
Srinivasan Sridhar, *Rhône-Poulenc, Inc.*
Johnny O. Wright, *Amoco Corporation*
A. Sumner West, *CCPS Staff Consultant*

This subcommittee prepared the broad outline for the book, identified the scope and major key references, and selected the title "Guidelines for Chemical Reactivity Evaluation and Application to Process Design" as representative of the concepts desired. The TNO Prins Maurits Laboratory, Rijswijk, The Netherlands, was chosen as the contractor with Dr. A. Henk Heemskerk as the project manager.

The subcommittee provided guidance and fruitful input to the contractor during the preparation of this book, and served as principal editors of the final draft received from TNO Prins Maurits Laboratory.

ACKNOWLEDGMENTS

The Center for Chemical Process Safety expresses sincere appreciation to the members of the staff of TNO Prins Maurits Laboratory, Rijswijk, The Netherlands, who prepared this document. Special recognition is given to the following staff members:

Project Manager: A. Henk Heemskerk
Principal Authors: A. Henk Heemskerk
                          Aat C. Hordijk
                          André T. Lanning
                          Johan C. Lont
                          Hans Schell
                          Peter Schuurman

The Center for Chemical Process Safety thanks all of the members of the Reactive Chemicals Subcommittee listed in the Preface for providing technical guidance and significant editing effort in the preparation of this book. Appreciation is also expressed to the employers of the subcommittee members for providing the time to work on this project.

The advice and support of the CCPS Technical Steering Committee is acknowledged.

GLOSSARY

Activation energy: the constant E_a in the exponential part of the Arrhenius equation associated with the minimum energy difference between the reactants and an activated complex (transition state), which has a structure intermediate to those of the reactants and the products, or with the minimum collision energy between molecules that is required to enable a reaction to take place; it is a constant that defines the effect of temperature on reaction rate.

Adiabatic: a system condition in which no heat is exchanged between the system and its surroundings; in practice, near adiabatic conditions are reached through good insulation.

Adiabatic induction time: the delay time to an event (spontaneous ignition, explosion, etc.) under adiabatic conditions starting at operating conditions.

Adiabatic temperature rise: maximum temperature increase, readily calculated, that can be achieved; this increase would occur only when the substance or reaction mixture decomposes completely under adiabatic conditions.

Apparent activation energy: in this book, the constant E_a that defines the effect of temperature on the global reaction rate.

Arrhenius equation: the equation is $k = A \exp(-E_a/RT)$, where k is the reaction rate constant; the pre-exponential factor A and the activation energy E_a are approximately constant for simple reactions.

Arrhenius plot: plot of the logarithm of the reaction rate constant k versus the reciprocal of the absolute temperature T; the plot is a straight line with slope of $-E_a/R$ for uncomplicated reactions without autocatalysis or inhibitor depletion effects.

Autocatalytic reaction: reaction in which the rate is increased by the presence of one or more of its intermediates and/or products.

Autoignition temperature: the minimum temperature required to initiate or cause self-sustained combustion of a substance in air with no other source of ignition; the autoignition temperature is not a material-intrinsic property and therefore depends on the conditions of measurement.

Batch reactor: reactor in which all reactants and solvents are introduced prior to setting the operating conditions (e.g., temperature and pressure).

Bench scale: operations carried out on a scale that can be run on a laboratory bench.

BLEVE (Boiling-Liquid-Expanding-Vapor-Explosion): a type of rapid phase transition in which a liquid contained above its atmospheric boiling point is rapidly depressurized, causing a nearly instantaneous transition from liquid to vapor with a corresponding energy release; a BLEVE is often accompanied by a large fireball when a flammable liquid is involved since an external fire impinging on the vapor space of a pressure vessel is a common BLEVE scenario; however, it is not necessary for the liquid to be flammable for the occurrence of a BLEVE.

Blowdown: rapid discharge of the contents of a vessel; also, a purge stream as from boiler water.

Condensed phase explosion: an explosion of a liquid or solid substance.

Confined explosion: an explosion that starts inside a closed system (e.g., vessel or building).

Containment: a system in which no reactants or products are exchanged between the chemical system and its surroundings (closed system).

Continuous reactor: a reactor characterized by a continuous flow of reactants into and a continuous flow of products from the reaction system; examples are the plug flow reactor (PFR) and the continuous stirred tank reactor (CSTR).

Continuous stirred tank reactor (CSTR): an agitated tank reactor with a continuous flow of reactants into and products from the agitated reactor system; ideally, composition and temperature of the reaction mass is at all times identical to the composition and temperature of the product stream.

Critical coolant temperature: the maximum temperature of coolant, either gas or liquid, at which all heat generated by a chemical reaction can still be transferred to the coolant.

Critical mass: the minimum mass required to enable the occurrence of an explosion under specified conditions.

Critical steady-state temperature (CSST): the highest ambient temperature at which self-heating of a material as handled (in a package, container, tank, etc.) does not result in a runaway but remains in a stationary condition (*see* Self-Accelerating Decomposition Temperature).

Decomposition energy: the maximum amount of energy which can be released upon decomposition.

Decomposition temperature: temperature at which decomposition of a substance occurs in a designated system; it depends not only on the identity of the substance but also on the rate of heat gain or loss in the system.

GLOSSARY xxi

Defensive measures: measures taken to reduce or mitigate the consequences of a runaway to an acceptable level.

Deflagration: a release of energy caused by a rapid chemical reaction in which the reaction front propagates by thermal energy transfer at subsonic speed.

Design Institute for Emergency Relief Systems (DIERS): organization of the American Institute of Chemical Engineers to investigate and report on design requirements for vent systems for a variety of circumstances.

Detonation: a release of energy caused by an extremely rapid chemical reaction of a substance in which the reaction front propagates by a shock wave at supersonic speed.

Differential scanning calorimetry (DSC): a technique in which the difference of energy inputs required to keep a substance and a reference material at the same temperature is measured as a function of temperature, while the substance and the reference material are subjected to a controlled temperature program.

Differential thermal analysis (DTA): a technique in which the temperature difference between a substance and a reference material is measured as a function of temperature, while the substance and the reference material are subjected to a controlled temperature program.

Endothermic reaction: a reaction is endothermic if energy is absorbed; the enthalpy change for an endothermic reaction is a positive value.

Enthalpy of reaction: the net difference in the enthalpies of formation of all of the products and the enthalpies of all of the reactants; heat is released if the net difference is negative.

Event tree (analysis): a graphical logic diagram which identifies and sometimes quantifies the frequencies of possible outcomes following an initiating event.

Exothermic reaction: a reaction is exothermic if energy is released; the enthalpy change for an exothermic reaction is a negative value.

Fault tree (analysis): a method for the logical estimation of the many contributing failures that might lead to a particular outcome (top event).

Failure Mode Effect (and Criticality) Analysis [FME(C)A]: a technique in which all known failure modes of components or features of a system are considered in turn and undesired outcomes are noted; a criticality ranking of equipment may also be estimated.

Hazard: a chemical or physical condition that has the potential for causing harm or damage to people, property, or the environment.

Hazard and Operability Study (HAZOP): a systematic, qualitative technique to identify process hazards and potential operating problems using a series of guide words to generate process deviations.

Hazardous chemical reactivity: property of a chemical substance that can react yielding increases in temperature and/or pressure too large to be absorbed by the environment surrounding the system.

Incident: an unplanned event or series of events and circumstances that may result in an undesirable consequence.

Inherently safe: maintenance of a system in a non-hazardous state after the occurrence of any credible worst case deviations from normal operating conditions.

Isoperibolic system: a system in which the controlling external temperature is kept constant.

Isothermal: a system condition in which the temperature remains constant; this implies that heat internally generated or absorbed is quickly compensated for by sufficient heat exchange with the surroundings of the system.

Kinetic data: data that describe the rate of change of concentrations, heat, pressure, volume, etc. in a reacting system.

Law of Conservation of Energy: energy can change only in form, but never be lost or created.

Loop reactors: continuous flow reactors in which all or part of the product stream is recirculated to the reactor, either directly or mixed with a reactant supply stream.

Maximum pressure after decomposition: the maximum pressure obtainable in a closed vessel; this pressure is a function of the adiabatic temperature rise and the specific gas production.

Microcalorimetry: essentially isothermal techniques of high sensitivity in which very small heat fluxes from the reacting materials are measured; differential microcalorimetry is a technique to determine heat fluxes from the reacting materials compared with those of a reference material.

Onset temperature: temperature at which a detectable temperature increase is first observed due to a chemical reaction; it depends entirely on the detection sensitivity of the specific system involved; scale-up of onset temperatures and application of rules-of-thumb concerning onset temperatures are subject to many errors.

Overadiabatic mode: a quasi-adiabatic mode in which the (small) energy leaks to the environment are overcompensated for by input of supplementary energy.

Phi-factor: a correction factor which is based on the ratio of the total heat capacity of a vessel and contents to the heat capacity of the contents; the Phi-factor approaches one for large vessels.

Plug flow reactor (PFR): a tube reactor in which the reactants are fed continuously at one end and the products are removed continuously from the other end; concentration and heat generation change along the length of the tube; the PFR is often used for potentially hazardous reactions because of the relatively small inventory in the system.

GLOSSARY   xxiii

Pre-exponential factor: the constant A in the Arrhenius equation (also called the frequency factor); this pre-exponential factor is associated with the frequency of collosions between molecules, and with the probability that these conditions result in a reaction (*see also* Activation Energy and Arrhenius Equation).

Preventive measures: measures taken at the initial stages of a runaway to avoid further development of the runaway or to reduce and mitigate its final effects.

Quasi-adiabatic: a vessel condition that allows for small amounts of heat exchange; this condition is typical in testing self-heating by oxidation that is characterized by gas flows (although well-controlled in temperature) into and/or out of the test vessel; this condition is typical as well in tests where heat transfer is avoided by active control, that is, the ambient temperature is kept identical to the test vessel temperature, such that an adiabatic condition is approached.

Quenching: Abruptly stopping a reaction by severe cooling or by catalyst inactivation in a very short time period; used to stop continuing reactions in a process thus preventing further decomposition or runaway.

Rate of reaction: technically, the rate at which conversion of the reactants takes place; the rate of reaction is a function of the concentrations and the reaction rate constant; in practical terms, it is an ambiguous expression that can describe the rate of disappearance of reactants, the rate of production of products, the rate of change of concentration of a component, or the rate of change of mass of a component; units are essential to define the specific rate of interest.

Reaction: the process in which chemicals/materials (reactants) are converted to other chemicals/materials (products); types of reactions are often characterized individually (e.g., decompositIons, oxidations, chlorinations, polymerizations).

Reaction kinetics: a mathematical description of reaction rates in terms of concentrations, temperatures, pressures, and volumes that determine the path of the reaction.

Reaction rate constant: the constant in the rate of reaction equation; it is a function of temperature as represented in the Arrhenius equation.

Reflux: a system condition in which a component in the reaction system (usually a solvent or diluent) is continuously boiled off, condensed in a nearby condenser, and then returned to the reaction system; reflux is often used to operate at a preset temperature or to avoid operating at unacceptably high temperatures.

Risk: a measure of potential economic loss, environmental damage, or human injury in terms of both the probability of the loss, damage, or injury occurring and the magnitude of the loss, damage, or injury if it does occur

Runaway: a thermally unstable reaction system which shows an accelerating increase of temperature and reaction rate which may result in an explosion; three stages can be identified as: (1) a first stage in which the temperature increases slowly and essentially no gases are generated, (2) a second stage in which gas generation starts to occur and thermal gradients may occur depending on the rate of agitation and on the physical characteristics of the reaction system, and (3) a third stage in which a rapid increase in temperature and reaction rate occur, usually accompanied by temperature gradients and significant pressure increases.

Selectivity: the ratio of the amount of a desired product obtained to the amount of a key reactant converted.

Self-Accelerating Decomposition Temperature (SADT): the lowest ambient temperature at which a runaway decomposition is observed within seven days; the test is run with unstable substances, such as a peroxide, in its commercial shipping container, and the reported result applies only for the container used.

Semi-Batch Reactor (SBR): a type of batch reactor from which at least one reactant is withheld and then added at a controlled rate, usually to control the rate of heat generation or gas evolution; both heat generation and concentrations vary during the reaction process; products are removed from the reactor only upon conclusion of the reaction process.

Stationary conditions: conditions that are characterized by constant concentrations and temperatures as a function of time (i.e., the time derivatives are zero).

Thermally unstable: chemicals and materials are thermally unstable if they decompose or degrade as a function of temperature and time within a credible temperature range of interest.

Time to maximum reaction rate: the measured time to the maximum reaction rate during a runaway or rapid decomposition; the specific result is highly contingent on the test method used.

Top event: the unwanted event or incident at the "top" of a fault tree that is traced downward to more basic failures using logic gates to determine its causes and likelihood of occurrence.

Unconfined vapor cloud explosion: explosive oxidation of a flammable vapor cloud in a nonconfined space (e.g., not in vessels or buildings); the flame speed may accelerate to high velocities and can produce significant blast overpressures, particularly in densely packed plant areas.

Unstable substance/material: substance or material that decomposes, whether violently or not, in the pure state or in the state as normally produced.

Venting: an emergency flow of vessel contents out of the vessel thus reducing the pressure and avoiding destruction of the unit from over-pressuring; the vent flow can be single or multiphase, each of which results in different flow and pressure characteristics.

LIST OF SYMBOLS

A	pre-exponential factor (Arrhenius equation)
A_p	peak area, m^2
A_s	surface area, m^2
C_p	specific heat at constant pressure, J/(kg °C)
C_v	specific heat at constant volume, J/(kg °C)
C_{ves}	specific heat of vessel, kJ/°C
c	concentration, kg/m^3
c_R	reactant concentration, mols/unit volume
d	diameter or thickness, m
dp/dt	rate of pressure change, bar/s
dT/dt	rate of temperature change, °C/s
E_a	activation energy, J/mol
F	frequency of incidents
F	specific energy (= force constant), kJ/kg
FF	fouling factor (heat transfer), J/(m^2 s °C)
G	gas flow, m^3/s
G	Gibbs free energy
H	enthalpy, kJ/kg
h	film heat transfer coefficient, J/(m^2 s °C)
K	constant
k	reaction rate constant
l	typical length, m
MW	molecular weight
m	mass, kg
N	number of atoms in a molecule
N_{Bi}	Biot number, $(hx)/\lambda$
N_{Nu}	Nusselt number, $(hd)/\lambda$
N_o	number of moles of oxidant
N_{Pr}	Prandtl number, $(C_p\mu)/\lambda$
N_{Re}	Reynolds number, $(d^2 N_s \rho)/\mu$

N_s	rotational speed, revolutions/minute
n	number of incidents
OB	oxygen balance
p	pressure, bar
Q	quantity of heat, J, or energy per unit mass, J/kg, or energy per unit mass per time, J/(kg s)
q	rate of heat generation, J/s
R	molar gas constant, kJ/(kmol °C)
r	reaction rate, mol/(m^3 s)
S	entropy, kJ/(kmol °K)
S	surface area, m^2 ($\approx A_s$)
T	temperature, °C
T_c	critical ambient or critical coolant temperature, °C
T_j	jacket temperature, °C
T_m	temperature of heating/cooling medium, °C
T_{nr}	temperature of no return, °C
T_o	onset temperature (initiation of reaction), °C
T_r	reaction temperature, °C
t	time, s
U	internal energy
U	overall heat transfer coefficient, J/(m^2 s °C)
U_f	internal energy of formation
V	volume, m^3
V_b	volume of autoclave, m^3
x	radius or dimension, m
ΔG_r	Gibbs free energy of reaction
ΔH_c	enthalpy of combustion (complete), J/kg
ΔH_d	enthalpy of decomposition, J/kg
ΔH_f	enthalpy of formation, J/mol
ΔH_o	enthalpy of oxidation, J/kg
ΔH_r	enthalpy of reaction, J/mol
ΔH_v	enthalpy of evaporation, J/kg
ΔS	change in entropy, kJ/(kmol °K)
ΔS_r	change in reaction entropy, kJ/(kmol °K)
ΔT_{ad}	adiabatic temperature rise, °C
ΔT_{lm}	logarithmic temperature difference, °C $= (\Delta T_{in} - \Delta T_{out})/(\ln \Delta T_{in} - \ln \Delta T_{out})$
ΔU_r	internal energy of reaction, J/mol
ΔV	volume change, m^3
ΔV_r	reaction volume change, m^3
δ	ratio of heat production rate to heat removal rate

LIST OF SYMBOLS

Θ	shape factor
λ	thermal conductivity coefficient, J/(m s °C)
ρ	density, kg/m³
σ_p	selectivity
τ_i	adiabatic induction time, minutes
Φ	Phi-factor (thermal inertia)
Ψ	mass flow, kg/s

Subscripts not otherwise indicated:

0	parameter value at $t = 0$
A, B,...	reactant/product identification
m	heating/cooling medium
max	maximum
p	process
ref	reflux
s	surface

1
INTRODUCTION

1.1 GENERAL

The intent of this Guidelines book is to provide the principles for the evaluation of chemical reactivity and for use of this information to design and operate safer chemical plant processes. Special emphasis is placed on the use of state-of-the-art methodology in the areas of theory, testing methods, and applications in design and operation of inherently safer processes.

This book presents significant guidelines to aid in avoiding runaway reactions. These reactions can be the cause of catastrophic events because the sudden energy release can cause damage and injury from direct effects of high temperatures and pressures and can cause illness and death from the release of toxic materials. The responsibility of chemists, chemical engineers, and others in preventing these events is to be knowledgeable in and to understand the reactivities of the materials involved, and to apply this knowledge effectively in the design, operation, and maintenance of chemical processes.

The book is directed to those persons involved in research, process development, pilot plant scale-up, process design, and commercial plant operations. It is important for technical people considering alternative process routes to know the potential hazards from the main reactions and from the unwanted side reactions in each case so that the hazards of reactivity are included in the factors reviewed in developing and selecting the final process route.

Heat evolution calculations and laboratory testing are usually needed to define the reactivity hazards. This book outlines methods for identifying hazardous reactions and determining safe conditions. Data are needed on various rate phenomena, enthalpies, and other thermal properties.

The information in this book is concerned primarily with **prevention** of runaway reactions rather than mitigation effects after such events have occurred. Further, this book covers technical issues and not specifically management techniques.

The following classes of materials are included:

1. self-reacting chemicals (including those that deflagrate or detonate without the presence of oxygen),
2. chemicals that react with common contaminants such as water, oxygen, and sunlight, including substances that form peroxides, and
3. pyrophoric substances.

Specifically excluded from the guidelines are the following topics, which are or will be covered in other AIChE/CCPS publications:

1. dusts and dust clouds,
2. vapor cloud issues,
3. vapor phase fires and explosions, and
4. transportation issues.

The guidelines describe the general approach to safer process design and operation using basic principles of thermodynamics, chemical kinetics, and reaction engineering. Included are some general reaction engineering concepts that can contribute to the design of safer chemical processes. Emphasis is placed on the need to evaluate process safety at an early stage by the process development team. A recurrent theme in the application section of the book is that a safe process is as important a goal as a more economic or productive process.

The definition of reactive chemical suggested by Kohlbrand [1] was useful in determining the content of these guidelines:

> Hazardous chemical reactivity is any chemical reaction with the potential to exhibit rates of increase in temperature and/or pressure too high to be absorbed by the environment surrounding the system. Included are both reactive materials (those which enter into a chemical reaction with other stable or unstable materials) and unstable materials (those which in a pure state or as normally produced decompose or undergo violent changes).

There are three main parameters tha determine the design of safe chemical processes: (1) the potential energy of the chemicals involved, (2) the rates of their potential reactions and/or decompositions, and (3) the process equipment. This is illustrated as the triangle in Figure 1.1.

The first key factor, energy, is involved in the production of any chemical. Design of a safe process requires an understanding of the inherent energy (exothermic release/endothermic absorption) during chemical reactions. This information can come from the literature, from thermochemical calculations, or from proper use of testing equipment and procedures. The potential pressure that may be developed in the process is also a very important design consideration.

The second key process design parameter is the reaction rate, which depends on temperature, pressure, and concentrations. Rates of reaction

1.1 GENERAL

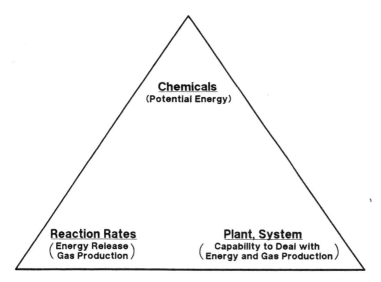

FIGURE 1.1. Key Parameters That Determine Design of Safe Chemical Plants.

during normal and abnormal operation (including the worst credible case) must be determined in order to design inherently safe processes.

Plant process and equipment design are elements of the third key parameter. Any heat that is generated by the reaction must be removed adequately, and any gas production must be managed. The effects and requirements of scale-up (that is, the relation between bench-scale and plant equipment) must be considered.

These three parameters interact. For example, a large amount of potential energy can be removed during normal operation if the rate of energy release is relatively small and is controlled by sufficient cooling capacity of the plant unit. However, if the cooling capacity of the plant unit appears insufficient because of the rate of energy release, a hazard assessment can be used to determine the necessary cooling design requirements for the operation.

In most cases, data that are obtained through theoretical approaches (literature, data bases, software programs) may not be sufficient for final plant design. Experimental work is usually required on various scales depending on the extent of reactivity. Therefore, the application of well designed experimental test methods is of prime importance to define hazardous conditions. Numerous test methods are available using a variety of sample sizes and conditions.

Identification of hazardous chemicals through thermodynamic and kinetic analyses is discussed in Chapter 2. This hazard identification makes use of thermal analysis and reaction calorimetry. In Chapter 2, an overview of the theory of thermodynamics, which determines the reaction (decomposition) phenomena is presented, including calculation methods. Experimental methods are evaluated to determine the initiation of a runaway and to determine the effect of decompositions that may occur on runaway. The aspects of stability, compatibility, catalyzing behavior, and reducing and oxidizing phenomena are also treated in this chapter.

In Chapter 3, the reaction system is discussed using the heat and mass balances, and interaction with the equipment. Scale-up affects both temperature and pressure profiles, which vary with types of reactor systems and sizes. Relevant test methods for scale-up and for process design are covered, including discussions on the methods as well as the relative advantages and disadvantages. Typical approaches for safe design and for defensive measures are presented. The theoretical and experimental subjects in Chapters 2 and 3 are illustrated by the use of examples.

In addition to the evaluation of chemical process hazards, and the proper applications of the evaluation to process design and operation, the management systems are important to assure operation of the facilities as intended. Brief introductions into hazard identification and quantification, and into management controls from the perspective of process safety are presented in Chapter 4. Future trends are also briefly reviewed here.

Extensive discussions of hazard evaluation and quantification are covered in the AIChE/CCPS *Guidelines for Hazard Evaluation Procedures* [2, 3] and *Guidelines for Chemical Process Quantitative Risk Analysis* [4]. Management control is extensively treated in the AIChE/CCPS *Guidelines for Technical Management of Chemical Process Safety* [5], *Plant Guidelines for Technical Management of Process Safety* [6], and the *Guidelines for Auditing Process Safety Management Systems* [7]. General design considerations for process safety are covered in AIChE/CCPS *Guidelines for Engineering Design for Process Safety* [8]. A summary of the CCPS program on reactive chemicals also has been published [9].

1.2 CHEMICAL REACTIVITY

In the process industries, chemicals are converted into other chemicals in a well-defined and well-controlled manner. Uncontrolled chemical reactions occur under abnormal conditions, for example, malfunctioning of the cooling system and incorrect charging. Temperature, pressure, radiation, catalysts, and contaminants such as water, oxygen from air, and equipment lubricants can influence the conditions under which the reactions (controlled and uncontrolled) take place.

The rate at which a chemical is converted is an exponential function of temperature. In comparing reaction rates among chemicals at a certain temperature, some chemicals show a high stability and others a relatively low stability.

Almost all reactions show a heat effect. When heat is produced during a reaction (exothermic), a hazardous situation may occur depending on the reaction rate, the quantity of heat that is generated, the capacity of the equipment to remove the heat, and the amount of gas produced during the reaction.

Although thermal decomposition (and runaway) is often identified with the inherent reactivities of the chemicals involved, it must be emphasized that hazards can arise from induced reactions as discussed in Chapter 2. These induced reactions may be initiated by heat, contamination, or mechanical means (e.g., shock, friction, electrostatic spark).

1.3 DETONATIONS, DEFLAGRATIONS, AND RUNAWAYS

An explosion is a rapid expansion of gases resulting in a rapidly moving pressure or shock wave. The expansion can be the result of a rapid chemical reaction. If the front velocity of the shock wave exceeds the speed of sound in the material, the energy is transferred by shock compression resulting in what is termed a detonation. At front velocities lower than the speed of sound, the energy is transferred by heat resulting in what is termed a deflagration.

The effect of a **detonation** depends on the shock wave, that is, an immediate peak overpressure followed by a longer period with an underpressure. The strength of the shock wave depends on the mass of the detonating materials. Detonations are mostly induced by initiation sources. In some cases, a deflagration may make a transition into a detonation. Working with chemicals and systems under plant conditions where a detonation can be induced is **NOT** recommended. Whether or not a chemical or chemical system can detonate can be determined only by specific tests as outlined in Chapter 2.

The effect of a **deflagration** depends on the rapid energy release in the form of heat, and on the pressure increase coinciding with the deflagration. The effect of a deflagration cannot be determined on a theoretical basis. A decomposition rate is far higher than would be expected on the basis of kinetic data. The tests by which deflagration behavior can be investigated are described in Chapter 2. Both preventive and defensive measures must be considered in dealing with a deflagration.

A **runaway reaction** proceeds by a general temperature increase because of insufficient heat removal. This type of runaway is generally encountered in large units, including storage vessels, and in well-stirred systems. A runaway may be caused by a rapid decomposition or oxidation reactions in units other than reactors. In a reactor, various phenomena may cause a runaway, including accumulation and/or mischarging of reactants, incorrect handling of catalysts, cooling problems, or loss of agitation.

In most cases, a thermal runaway depends on the balance between heat generation and heat removal. When heat removal is insufficient, the temperature will increase according to the reaction kinetics. Gases may either be formed as products of the reaction or, in later stages, as decomposition products at the elevated temperatures encountered. In general, there are two alternatives available to handle the gas production. Either the vessel must be designed to withstand the total pressure involved, or a vent system must be designed so that the vessel pressure never exceeds the design pressure during the runaway. In case of a thermal runaway, the use of preventive measures is recommended.

1.4 ASSESSMENT AND TESTING STRATEGIES

Recommended testing procedures depend on the stage of development of the process as indicated in Table 1.1. During early developmental chemistry work, only small amounts of materials will be available. In many cases, only theoretical information from the literature or from calculations is readily available. Screening tests can be run to identify the reaction hazards. Also, data for pilot plant considerations can be obtained.

In the pilot plant stage, additional material becomes available so that the reaction hazards can be investigated more extensively. Process control features and deviations from normal operating conditions can be checked. Operating procedures can be drafted and checked. Emergency procedures can be defined.

TABLE 1.1
Suggested Stages in Assessment of Reactivity by Scale

Stages	Aspect
1. Development Chemistry—Characterization of materials	Characterization of process alternatives Choice of process Suitability of process Screening for chemical reaction hazards
2. Pilot Plant—Chemical reaction hazards	Influence of plant selection on hazards Definition of safe procedures Effects of expected variations in process conditions Definition of critical limits
3. Full Scale Production—Reevaluation of chemical reaction hazards	Newly revealed reactivity hazards from plant operations Management of changes Update of safety procedures as required Ongoing interaction of process safety with engineering, production, economic, andcommercial aspects of the process

1.4 ASSESSMENT AND TESTING STRATEGIES

During full scale production, particularly initially, chemical reaction hazards may be reevaluated. More tests may be necessary as a consequence of increased knowledge of the process, changed production requirements, or other process changes such as the use of different feed stocks.

A typical chronology for testing is shown on Table 1.2. The tests provide either qualitative or quantitative data on onset temperature, reaction enthalpy, instantaneous heat production as a function of temperature, maximum temperature, and/or pressure excursions as a consequence of a runaway, and

TABLE 1.2
Typical Testing Procedures by Chronology

Subject	Property to Be Investigated	Typical Instrument Information
Identification of exothermic activity	Thermal stability	DSC/DTA
Explosibility of individual substances	Detonation Deflagration	Chemical structure Tube test Card gap Dropweight Oxygen balance High rate test Explosibility tests
Compatibility	Reaction with common contaminants (e.g., water)	Specialized tests
Normal reaction	Reaction profile Effect of change Gas evolution	Bench-scale reactors (e.g., RC1)
Minimum exothermic runaway temperature	Establish minimum temperature	Adiabatic Dewar Adiabatic calorimetry ARC
Consequence of runaway reaction	Temperature rise rates Gas evolution rates	Adiabatic Dewar Adiabatic calorimetry Pressure ARC VSP/RSST RC1 pressure vessel

ARC = Accelerating Rate Calorimeter (Columbia Scientific Instrument Corp.)
DSC = Differential Scanning Calorimeter
DTA = Differential Thermal Analysis
RC1 = Reactor Calorimeter (Mettler-Toledo Inc.)
RSST = Reactive System Screening Tool (Fauske and Associates)
VSP = Vent Size Package (Fauske and Associates)

additional data useful for process design and operation. The test equipment is discussed in both Chapters 2 and 3.

A detailed strategy for the approach to safety testing is provided in Chapter 2 (Figures 2.3, 2.4, and 2.5) and in Chapter 3 (Figure 3.4). These schemes are directed to the investigation of thermal instabilities, chemical incompatibilities, including acid, water, and oxygen incompatibility, and other factors important to potential unstable behavior.

2

IDENTIFICATION OF HAZARDOUS CHEMICAL REACTIVITY

2.1. SUMMARY/STRATEGY

2.1.1 Introduction

The main subject of this chapter is the identification of hazardous chemicals, materials, mixtures, and reaction masses. The chapter deals with undesired decompositions and hazardous reactions. A basic knowledge of the chemistry involved, and, in particular, with the thermodynamics and kinetics, is required. Furthermore, it is important to have a test strategy to recognize and assess the hazards associated with the energetic materials, mixtures, and reaction masses.

2.1.2 Hazard Identification Strategy

Figure 2.1 presents a flow chart that outlines a plan for the initial theoretical hazard evaluation of substances and reaction masses. This approach may be applied to evaluate the potential hazard of the substance on theoretical grounds provided that the molecular structure of the specific chemical is known.

Initially, the literature is searched for relevant data on the substance (physical–chemical properties, thermodynamics, incidents, case studies, and so forth). If insufficient data are available, the usual case, a systematic investigation procedure comprising three main subjects must be initiated for the material in question.

First, potentially unstable molecular groups in the molecule are identified (see Section 2.2.2).

Second, the potential energy and reactivity of the substance is determined. Two methods are applied here:

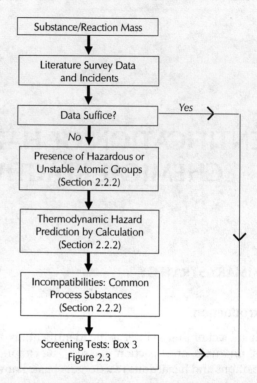

FIGURE 2.1. Initial Theoretical Hazard Identification Strategy.

1. The oxygen balance of the substance is calculated (see Section 2.2.3.1). This oxygen balance relates to the number of oxygen and reducing atoms in the substance itself. If all reducing atoms can be oxidized completely without an excess of oxygen (i.e., a stoichiometric ratio), the oxygen balance is zero, and the energy generation of the substance is maximum and is independent of the external oxygen concentration.
2. The heat of decomposition and/or reaction (in absence of ambient oxygen) is calculated (see Section 2.2.3.2). If the value of the oxygen balance is less than −240 or higher than +160 (Table 2.12) *and* the calculated heat of reaction/decomposition is less than 100 cal/g (420 J/g), the substance in its pure form is regarded as having a very low potential to produce a deflagration or detonation [10, 11].

In the third step, the chemical structure is used to determine if the substance is compatible with materials which are common to the process unit, such as air, water, oxidizers and combustibles, acids, alkalies, catalysts, trace metals, and process utilities (see Section 2.2.4). Even if the substance is considered to be a non-explosion hazard (both nonenergetic and compatible with the

2.1. SUMMARY/STRATEGY
11

"process-common" materials), it still should be subjected to screening tests to ensure the absence of instabilities and the potential of a runaway reaction. A complete schematic strategy of testing is shown later in Section 2.1.5 as Figure 2.3. The type of experimental work involved depends on the stage of the process development and on the type of potential hazard.

2.1.3 Exothermic Reactions

A reaction is exothermic if heat (energy) is generated. Reactions in which large quantities of heat or gas are released are potentially hazardous, particularly during fast decomposition and/or complete oxidations.

Exothermic reactions lead to a temperature rise in the material if the rate of heat generation exceeds the rate of heat removal from the material to its surroundings (for self-heating, see also Chapter 3). The reaction accelerates due to the increasing temperature and may result in a thermal runaway. The increase in temperature will be considerable if large quantities of heat are generated in a short time. Many organic compounds that decompose exothermically will liberate pressure-generating condensable and noncondensable gases at high temperatures.

In addition to thermal runaways, which result from more-or-less uniform self-heating throughout the material, highly exothermic decompositions can be induced by the point source input of external energy, for example, fire, hot spots, impact, electrical sparks, and friction. In such a case, the decomposition travels through the material by either a heat or a shock wave. Therefore, the maximum quantities of both energy and gas that are generated by the exothermic reaction are prime parameters in estimating the potential reactivity hazards of a substance. Furthermore, the rates of energy generation and gas production are of utmost importance [12].

Even relatively small amounts of exothermic reaction or decomposition may lead to the loss of quality and product, to the emission of gas, vessel pressurization, and/or environmental contamination. In the worst case, an uncontrolled decomposition may accelerate into an explosion.

The types of explosions that may occur depend on the confinement of the reactive material, its energy content, its kinetic parameters, and the mode of ignition (self-heating or induced by external energy input). Explosions are characterized as physical or chemical explosions, and as homogeneous or heterogeneous as described in Figure 2.2.

A physical explosion, for example, a boiler explosion, a pressure vessel failure, or a BLEVE (Boiling Liquid Expanding Vapor Explosion), is not necessarily caused by a chemical reaction. Chemical explosions are characterized as detonations, deflagrations, and thermal explosions. In the case of a detonation or deflagration (e.g., explosive burning), a reaction front is present that proceeds through the material. A detonation proceeds by a shock wave with a velocity exceeding the speed of sound in the unreacted material. A

2. IDENTIFICATION OF HAZARDOUS CHEMICAL REACTIVITY

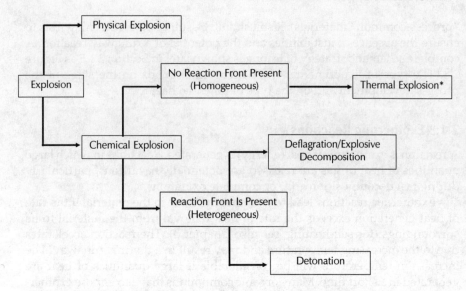

FIGURE 2.2. Types of Explosions. [*A: thermally driven (self-heating); B: chemically driven (e.g., inhibitor depletion, melting with decomposition, autocatalytic decomposition)].

deflagration proceeds by transport processes such as by heat (and mass) transfer from the reaction front to the unreacted material. The velocity of the reaction front of a deflagration is less than the velocity of sound in the unreacted material. Both types of explosions are often called heterogenous explosions because of the existence of a reaction front which separates completely reacted and unreacted material.

A thermal explosion is the third type of chemical explosion. In this case, no reaction front is present, and it is therefore called a homogenous explosion. Initially, the material has a uniform temperature distribution. If the temperature in the bulk material is sufficiently high so that the rate of heat generation from the reaction exceeds the heat removal, then self-heating begins. The bulk temperature will increase at an increasing rate, and local hot spots may develop as the thermal runaway proceeds. The runaway reaction can lead to overpressurization and possible explosive rupture of the vessel.

Explosion phenomena have occurred in all types of confined and unconfined units: reactors, separation and storage units, filter systems, pipe lines, and so forth. Typical reactions that may cause explosions are oxidations, decompositions, nitrations, and polymerizations. Examples of chemical and processing system characteristics that increase the potential for an explosion are the following:

- high decomposition or reaction energies,
- high rates of energy generation,
- insufficient heat removal (i.e., too large a quantity of the substances),

2.1. SUMMARY/STRATEGY

- the presence of an initiation source,
- substances with an oxygen balance close to zero,
- confinement, and
- large amounts and/or high rates of gas production.

2.1.4 Experimental Thermal and Reactivity Measurements

Experimental hazard evaluation includes thermal stability testing, solid flammability screening tests, explosibility testing, detailed thermal stability and runaway testing, and reactivity testing. Flammability testing of liquids, although highly important, is not within the scope of this book.

The recommended experimental evaluation is condensed in a number of flow charts following in which, in general, the most reliable and internationally recognized standard test methods are used [10, 12–22]. Details of the strategic testing scheme are covered in the following section.

2.1.5 Test Strategies

The potential thermal hazards associated with thermally unstable substances, mixtures, or reaction masses are identified and evaluated as in the flow charts Figures 2.3, 2.4, and 2.5. The potential hazards posed by reactivity—water reactivity, pyrophoricity, flammability, oxidizer contact, and so forth—are also included in Figure 2.3. The individual boxes in the flow charts are discussed below:

Screening Tests (Boxes 3 and 5)
In general, a theoretical evaluation of the hazardous chemical reactivity suggested in Box 2 is not sufficient by itself. The standard practice is to perform screening tests (Boxes 3 and 5).

The first aim of a thermal stability screening test (e.g., DSC/DTA) is to obtain data on the potential for exothermic decomposition and on the enthalpy of decomposition (ΔH_d). These data, together with the initial theoretical hazard evaluation, are used in reviewing the energetic properties of the substance (Box 4) and the detonation and deflagration hazards of the substance (Boxes 7 and 8). The screening tests also provide data on the thermal stability of the substance or mixture, on the runaway potential, on the oxidation properties, and to a lesser extent, on the kinetics of the reaction (Box 10).

The screening tests can be run in the absence or presence of air to differentiate the thermal hazards due to decomposition of the substance from those due to reactivity of the substance with oxygen.

Reliable and internationally accepted techniques for screening are differential scanning calorimetry (DSC) and differential thermal analysis (DTA). These techniques can provide exothermic enthalpy of reaction and observed

2. IDENTIFICATION OF HAZARDOUS CHEMICAL REACTIVITY

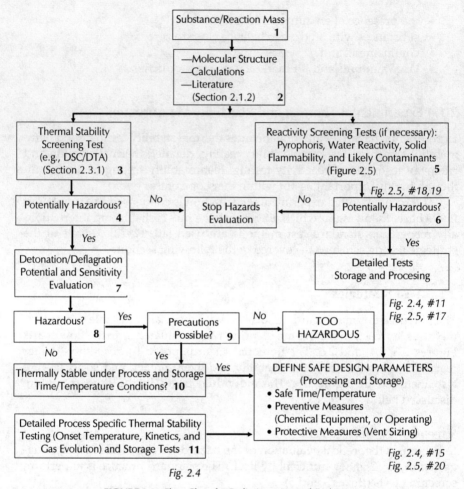

FIGURE 2.3. Flow Chart for Preliminary Hazard Evaluation

onset temperature. The results can be used to calculate *very approximate* reaction kinetic data (see Section 2.3.1.1). Flash point and combustibility testing are generally considered to be screening tests, but as stated previously, liquid/vapor flammability issues will not be discussed in this book.

Potentially Explosive (Box 4)

In Figure 2.3, the starting point (Box 2) is the compilation of the potential hazardous properties resulting from the theoretical evaluations. On the basis of this information, together with data obtained in the screening tests, it can be determined whether or not the substance is an energetic one. In general, a

2.1. SUMMARY/STRATEGY 15

substance has to be recognized as energetic ("yes" in Box 4) and thus potentially hazardous when:

1. the experimental enthalpy of decomposition (ΔH_d, in absence of air) is 50 to 70 cal/g (~200 to 300 J/g), noting that this range is highly dependent on the rate of reaction, rate of pressure increase, and process system design considerations, or
2. the structural formula contains hazardous molecular groups (see Table 2.5 for examples), or
3. an oxygen balance indicates explosive properties, or
4. a literature search reveals hazardous thermal properties.

If the substance does not meet any of the four mentioned criteria, the substance may be recognized as having a low risk to handle from the point of view of thermal hazards.

In the worst case, an enthalpy of decomposition of 50 to 70 cal/g results in an adiabatic temperature rise of approximately 100 to 200°C which is, as a rule-of-thumb, not regarded as critical under the condition that the substance does not easily produce a significant quantity of gas and thus, in a very general way, will not lead to a hazardous situation [23]. *However, this has to be evaluated for each individual process.*

Detonation and Deflagration (Box 7)

Generally, a substance is considered capable of detonating if it has a calculated enthalpy of decomposition in the absence of oxygen greater than 700 cal/g (~3000 J/g). Detonation tests (Box 7) are run to establish positively detonability and to measure appropriate properties. As discussed in Section 2.3.3.1, the likelihood of a detonation strongly depends upon the conditions of the testing (confinement, particle size, specific density, and so forth). Additionally with highly energetic substances, the sensitivity to friction and to impact must be determined (Section 2.3.3.3).

The velocity of propagation of a detonation (shockwave propagation) throughout the substance varies from 1000 to 6000 m/s [24] and can lead to complete fragmentation of the containment unit. If the substance is able to detonate and is sensitive to mechanical shock/friction, it is recognized as extremely hazardous! It may well be too hazardous to consider use in a plant process situation. In some cases, use of the substance is acceptable by blending with inerts, or by adequately protecting the surroundings. Mixing with inerts decreases the sensitivity and in many cases even excludes the possibility of a detonation in the system. The enthalpy of reaction in the presence of atmospheric oxygen—that is, the enthalpy of combustion, ΔH_c—is not relevant for reviewing the potential for detonability.

A minimum theoretical ΔH_d value of 700 cal/g is generally accepted as a criterion to indicate detonation potential. In CHETAH [12, 25–27] (see Section 2.2.3.3), a value of ΔH_d 700 cal/g is classified as "high" related to energy hazard potential. However, this is an approximate value, and several exceptions are known. For example, initiating explosives such as azides, which have a ΔH_d lower than 475 cal/g [24], are able to detonate and are very sensitive to mechanical shock and friction. Ammonium nitrate in some formulations and a number of organic peroxides are also able to detonate with ΔH_d even lower than 475 cal/g. Furthermore, the ΔH_d as measured by screening tests will often be lower than the enthalpy of explosion. Conditions of the screening tests (temperature and pressure) compared with the explosion conditions are less extreme and, as a result, the decomposition products may not be the same and thus the ΔH_d may differ. Therefore, the experimental values must be critically reviewed. If there is any doubt about the detonative potential of the substance or mixture, detonation testing is required (see Section 2.3.3.1).

A substance or mixture is potentially capable of deflagration if it has a ΔH_d greater than 250 cal/g (~1000 J/g), a "yes" in Box 8. Deflagration tests (Box 7) then should be carried out as well as tests for sensitivity to impact and friction.

Propagation rates of a deflagration vary from very low (1 to 10 mm/min) to very high (10 to 1000 m/s).

In Box 7, the deflagration properties following forced initiation are tested (see Section 2.3.3.2) at the temperature and pressure of processing. Forced initiation means initiation by external stimuli, such as a hot spot or a flame. If, after forced initiation at process conditions, the substance deflagrates violently (propagation rates of 10 to 1000 m/s) and is sensitive to impact or friction, it is recognized as extremely hazardous. In general, depending upon the rate of deflagration, acceptable precautions include venting or inerting (i.e., dilution by the addition of solid or liquid inerts) to decrease the deflagration rate and thus the rate of gas production. Specialized tests are available to investigate this phenomenon further.

The rule-of-thumb value of ΔH_d of 250 cal/g (calculated maximum value) as a criterion for determining deflagration potential is an approximate value. A number of values, varying from 170 to 300 cal/g, are stated in the literature [10, 27–29] as criteria. This ΔH_d range is valid for most of the known substances that are able to deflagrate. If a substance is able to deflagrate very rapidly, it is possible that the deflagration will propagate into a detonation, which is called deflagration–detonation transition (DDT). If necessary, this potential phenomenon can be investigated by special tests (see Section 2.3.3.2).

Thermal Stability under Process and Storage Conditions (Box 11)
For large-scale processing and storage situations, it is necessary to establish the thermal stability characteristics in order to determine the safe operating

2.1. SUMMARY/STRATEGY

and storage conditions. The heat production and thermal kinetic parameters are necessary to verify that the cooling capacity (natural or forced) of the process or storage facility can cope with the heat production. These parameters are also useful in establishing the forced cooling capacity requirements. For substances in normal process equipment such as distillation columns, dryers, extruders, and mixing units, in addition to the thermal stability parameters, the mitigation issues such as vent design for a runaway are important. This requires data concerning gas evolution and mass flux. These matters are dealt with more extensively in Chapter 3.

All theoretical calculations and tests performed in earlier stages, as well as the relevant literature, provide the inputs to the determination of the necessary detailed thermal stability testing for the substance, mixture, or reaction mass.

The three essential questions to be answered in this step are the following:

1. Is the process/storage stage sufficiently thermally stable under normal and worst credible time/temperature/pressure/chemical conditions?
2. If not, what are the important characteristics of the thermal decomposition if it should occur, such as time to onset, enthalpy of decomposition, self-heat rate, rate of pressure rise, rate of temperature rise, moles of gas/mole of substance? and
3. What are the important venting characteristics of the material?

The tests available relative to the first two questions are discussed in Sections 2.3.1 and 2.3.2, while those for the third question are reviewed in Chapter 3 (e.g., RSST and VSP).

The applicability of small-scale tests must be considered in view of the quantities to be handled, and the handling conditions at plant scale. Moreover, catalytic and autocatalytic effects, and inhibitor depletion play important roles in thermal stability evaluation, but they are not measured accurately by dynamic (temperature-ramped) screening tests. If the screening tests show that a substance is thermally unstable at, or even substantially above, the temperature range of large scale handling, then further investigations using more sensitive and dedicated tests and equipment are recommended as absolutely necessary.

Detailed Tests for Stability (Boxes 13 and 14)

Figure 2.4, a continuation of Figure 2.3, is a flow chart for a strategy for stability testing. The main aim of these tests is the determination of the following parameters:

1. the precise thermal stability characteristics, including catalysis, autocatalysis, inhibitor depletion, and compatibility,
2. the operating temperature to avoid hazardous decompositions,

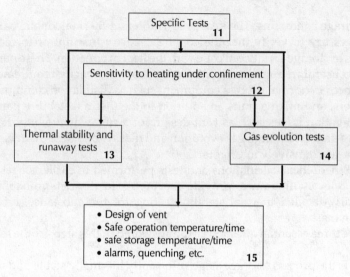

FIGURE 2.4. Flow Chart for a Strategy for Stability Testing

3. the time to maximum rate of reaction,
4. the adibatic temperature rise,
5. the gas evolution (mass flux) during decomposition and/or runaway reaction, and
6. the behavior of the material under external heat load.

The choice of test equipment to be used depends on the conditions, such as scale, temperature, mixing, and materials of construction, at which the substance or mixture is to be handled. The interpretation of the data from each of these tests is strongly dependent on the manner in which the test is run and on the inherent characteristics of the testing device. Guidance is provided along with each test description, particularly in the detailed sections later in this chapter.

The following equipment can be used to investigate the thermal stability, the operating temperature to avoid hazardous decomposition, the autocatalytic decomposition of a substance, inhibitor depletion in the system, and rates of pressure rise:

1. Accelerating rate calorimeter (ARC)—see Section 2.3.2.3,
2. Reactive system screening tool (RSST)—see Section 3.3.2.7,
3. Isothermal storage tests (IST), scanning or isothermal heat-flux microcalorimeters, thermal activity monitor (TAM)—see Section 2.3.2.1,
4. Dewar flask tests, adiabatic storage tests (AST)—see Section 2.3.2.2, and
5. Some types of isoperibolic (quasi-isothermal) equipment using relatively large quantities of sample—see Section 2.3.1.2.

2.1. SUMMARY/STRATEGY   19

These tests can also be used to evaluate the induction time for the start of an exothermic decomposition, and the compatibility with metals, additives, and contaminants. The initial part of the runaway behavior can also be investigated by Dewar tests and adiabatic storage tests. To record the complete runaway behavior and often the adibatic temperature rise, that is, the consequences of a runaway, the accelerating rate calorimeter (ARC) can be used, although it is a smaller scale test.

To investigate the gas evolution during decomposition and/or a runaway, both the ARC and RSST simultaneously record the temperature rise and the pressure rise, which is usually proportional to the gas evolution during decomposition.

Other types of equipment available to investigate the gas evolution are various autoclave tests (Section 2.3.3.2), isoperibolic autoclave tests (Section 2.3.1.2), and closed Dewar tests (Section 2.3.2.2). Mass flux data are also required in designing any vent facilities (Chapter 3).

Extrapolation of data from any and all of these tests to large scale must be made with care.

Reactive Substances (Boxes 17, 18, and 19)
A strategy for identifying the hazards of reactive substances is shown on Figure 2.5, which is a continuation of Figures 2.3 and 2.4.

For the purpose of hazard evaluation as outlined on Figure 2.5, reactive substances are defined as those that will react with other materials such as oxygen, water, and reducing and oxidizing agents (Section 2.2.4) in the processing or storage environment. In many cases, the structure of the substance or a literature review can be used to determine water reactivity, pyrophoric properties, and oxidizing properties. This flow chart does not really cover thermally unstable substances as such.

Substances, mixtures, or reaction masses are also evaluated and/or tested for compatibility with common process chemicals and contaminants (e.g., rust, water, air, heat transfer medium). Substances which are not energetic but show only decomposition and/or instability in the presence of oxygen (air), the so-called combustibles in Box 5 of Figure 2.3, can be included in the strategy shown on Figure 2.5.

Several tests have been developed to identify the hazards of reactive substances [10]. Test methods for determining pyrophoric properties, water reactivity, and oxidizing properties (Box 17) are discussed in Section 2.3.4.

Flammability testing of solids (Box 18) is discussed in Section 2.3.5. The reactivity of substances during large-scale processing and storage (Box 19), particularly of powders which may react with oxygen, is of the utmost importance because incorrect storage facilities may lead to self-heating, resulting in a smoldering fire. Furthermore, inhibitor depletion and autocatalytic effects may also play an important role in the stability of powders. The large-scale thermal stability of substances reacting with oxygen can be inves-

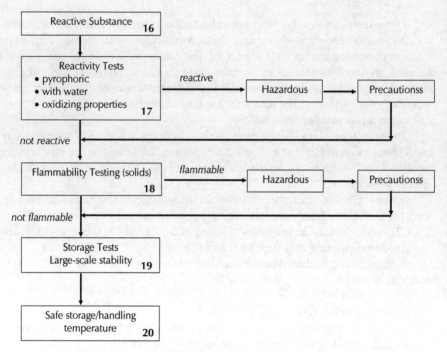

FIGURE 2.5. Flow Chart for Specific Experimental Hazard Evaluation for Reactive Substances

tigated by adiabatic storage tests (AST) or Dewar flask tests either with or without a supply of air (Section 2.3.2.2). Also, dedicated stability tests for powders are available (Section 2.3.2.4).

2.1.6 Overview of Thermal Stability Test Methods

An overview of typical calorimetric techniques indicating sensitivities, principal application areas, and the usual data acquired is shown on Table 2.1. A brief summary of advantages and disadvantages of the various tests is also given. The column "principal applications" indicates only the major applications of the respective techniques. In any of the tests listed, it is possible to obtain additional data or to use the test equipment for completely different hazard evaluations once the techniques are fully understood and the tests are run by fully qualified technical personnel. Testing techniques are discussed later in Section 2.3 on Practical Testing.

2.1. SUMMARY/STRATEGY

TABLE 2.1
Overview and Comparison of Calorimetric Techniques

Test Method	Section of Chapter 2	Typical Sample Mass (g)	Typical Sensitivity (W/kg)	Thermal Inertia: Phi-factor	Principal Application[a]	Data Acquired	A=Advantage D=Disadvantage[b]
DSC/ DTA	2.3.1.1	0.0005– 0.020	1–20	—	1	T_o, ΔH_o reaction kinetics	A: 1,2,3 D: 4,6
ARC	2.3.2.3	2–10	0.5	1.5–4.5	2,5,8,9	T_o, ΔH_o reaction kinetics, pressure data[c]	A: 1,2,3,4 D: 5
Dewar Tests, AST	2.3.2.2	200– 700	0.015	1.05–3.5	3,5,6,7,9	T_o, ΔH_o reaction kinetics, pressure data[c]	A: 4,5,6 D: 1,2,3
IST, TAM, Calvet	2.3.2.1	5–20	0.0001– 0.005	—	4,5,9	T_o, ΔH_o reaction kinetics, pressure data[c]	A: 4,6 D: 1,3,7
SEDEX SIKAREX	2.3.1.2	5–30	0.5	3.5	2,5	T_o, ΔH_o reaction kinetics, pressure data[c]	A: 6 D: 1,3,5

[a]Explanation of Principal Application Codes:
  1 = screening
  2 = thermal stability
  3 = sensitive thermal stability
  4 = very sensitive thermal stability
  5 = study autocatalysis, contaminations, inhibitor depletion
  6 = reaction due to oxidation
  7 = runaway behavior (initial phase)
  8 = complete runaway behavior and simultaneous pressure measurements
  9 = time to maximum rate of reaction

[b]Explanation of Advantages/Disadvantages Codes:
  Advantages
  1 = quick
  2 = small sample quantity required
  3 = wide temperature range covered
  4 = sensitivity to T_o
  5 = low Phi-factor
  6 = accurate overall kinetics

  Disadvantages
  1 = time consuming
  2 = large quantities required
  3 = restricted temperature range
  4 = insensitivity to T_o
  5 = medium to high Phi-factor
  6 = small sample quantities—hard to obtain a representative sample
  7 = more test runs required

[c]Pressure data are optional and can be obtained only in experiments performed in closed Dewar units or autoclaves.

2.1.7 Examples of Interpretation and Application of Test Data

As indicated in Section 2.1.6, every test has its own characteristics, which must be thoroughly understood to design experiments properly and to interpret the resulting data. It is essential to recognize that these tests are conducted on small scale samples under conditions that do not duplicate all aspects of plant conditions. Therefore, an expert should be consulted both before testing is initiated and again for interpretation and evaluation of the test results. Section 2.3 later discusses in detail the approach to practical testing, following Section 2.2 which reviews the technical aspects.

In the current section, a few typical examples and problems which arise in determining thermal reactivity hazards are discussed.

Example 1: Typical Outputs of Thermal Stability Test Methods
As discussed in detail later in Section 2.3, various techniques with different working principles are available to identify the thermal reactivity hazards of individual substances and reaction mixtures. Some examples are presented here.

Figure 2.6 shows typical curves recorded for exothermic decompositions by four different test methods.

Figure 2.6A shows the temperature history of a material when heated at a constant rate. The sample temperature is lower than that of the heating medium as the sample heats up with little reaction occurring. After attaining a temperature at which significant reaction takes place, the sample temperature exceeds that of the heating medium. The curve for sample temperature is similar to that for Curve C.

Figure 2.6B shows the heat evolution in isothermal tests of material decomposing in an autocatalytic mode at two temperatures. Typical curves for autocatalytic decomposition are also shown in Figure 2.8 discussed later in Example 3 [30].

Figure 2.6C shows the temperature difference between reference and sample as recorded by differential thermal analysis (DTA). Note also the similar differential scanning calorimeter (DSC) curve later in Figure 2.13.

Figure 2.6D shows the temperature curve from a typical "heat–wait–search" operation of adiabatic calorimetry. The sample was held adiabatically at three temperatures without detecting self-heating. At the fourth step, self-heating was detected and, after a wait of 20 minutes, a runaway occurred.

Example 2: Comparison of Onset Temperature as Detected by Various Instruments
An example of the differences in measurement of onset temperature, T_o, and heat of decomposition, ΔH_d, for t-butyl peroxybenzoate obtained from several types of equipment is shown on Table 2.2 [17, 31]. It is clear that the observed value of T_o, apart from the chemical properties of the sample, depends

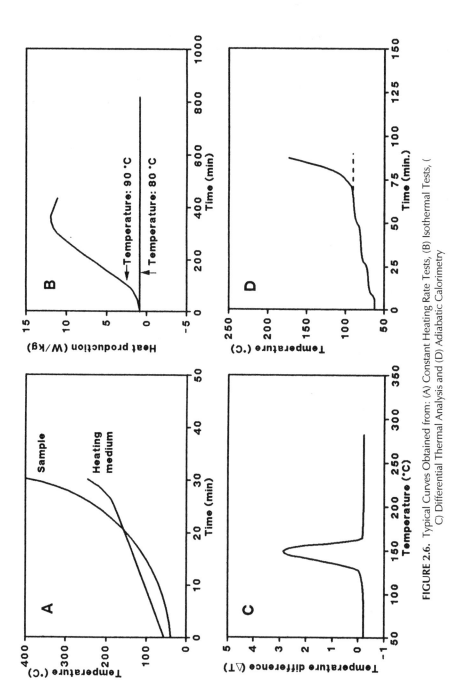

FIGURE 2.6. Typical Curves Obtained from: (A) Constant Heating Rate Tests, (B) Isothermal Tests, (C) Differential Thermal Analysis and (D) Adiabatic Calorimetry

TABLE 2.2
Comparison of T_o and ΔH_d for TBPB Using Different Calorimetric Techniques

Equipment	Experimental Parameters	Sample Mass	Recorded T_o (°C)	ΔH_d Recorded (J/g)	ΔH_d Corrected[a] (J/g)	Apparatus/ Comments
DSC	10°C/min	3.5 mg	122	1335	—	Sealed stainless steel pen
DSC	1°C/min	6.5 mg	93	1441	—	Sealed stainless steel pen
IST	Isothermal heat step 5°C	10 g	57[b]	—[c]	—	Glass vessel
AST	Start: 20°C, heat step 5°C, wait 24 hrs.	1000 g	55	—[c]	—	Glass Dewar vessel
ARC	Start: 50°C, heat step 5°C, wait 15 min, $\Phi = 2.32$	3.6 g	82	311	722	CSI—ARC "light" Hastelloy bomb
SEDEX	0.5°C/min "scanning"	5 g	84	—	—	Systag TSC 510/511 open tube
SIKAREX	0.125°C/min "scanning"	5 g	72	194	683	Systag Sikarex 3 open tube

[a] Corrected for thermal inertia [$\Delta H_d = \Phi \Delta H_d$ (recorded)].
[b] Isothermal temperature at which significant exothermic decomposition was detected.
[c] Incomplete conversion and thus no complete ΔH_d recorded.

strongly on the type of equipment used. This dependency is a result of, for example, the sensitivity of the apparatus, the heating rate (in case of a scanning technique), and the quantity of sample used.

In general, to establish T_o values for large scale purposes, tests discussed in Section 2.3.2 can be used. Under conditions of limited heat transfer, such as natural cooling in large storage vessels, relatively low heat production rates can lead to thermal runaways. Very sensitive test equipment is often needed to determine safe operating temperatures. Although an identical heat production rate can be readily controlled in an agitated system with circulating cooling, the decomposition will nonetheless result in loss of product quality.

As with the case presented in Table 2.2, values of ΔH_d determined in a closed cup DSC are reliable under most decomposition conditions. The ARC

2.1. SUMMARY/STRATEGY

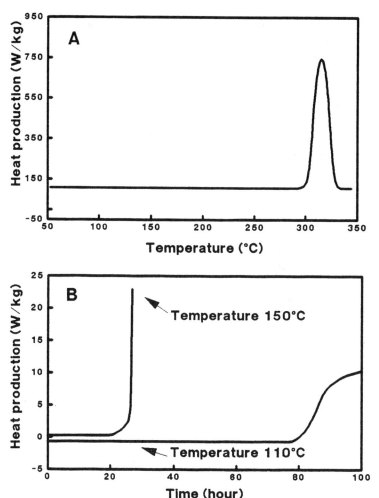

FIGURE 2.7. Depletion of Inhibitor Stability: DSC Curve (A) and Isothermal Curves (B) for an Inhibited Material.

can also be used for the determination of ΔH_d. In some cases, the corrected ΔH_d value will only be reliable when the self-heat rate of the sample can be matched by the calorimeter, a condition not often met during the final stages of energetic decompositions.

In all cases, care must be taken to identify secondary reactions which can occur at temperatures higher than those scanned and which can generate significant quantities of additional heat.

Example 3: Inhibitor Depletion/Autocatalytic Decomposition

The investigation of thermal hazards of materials containing an inhibitor, or which are sensitive to autocatalysis, is a difficult problem. Based on scanning types of screening tests, these materials may appear to be thermally stable at the process temperature, for example, as shown in Figure 2.7A indicating an exothermic decomposition with an onset temperature of 270°C. However, in a scanning experiment (e.g., a DSC), complete depletion of the inhibitor or generation of an amount of decomposition product which causes autocatalysis is reached only at higher temperatures, resulting in the observation of an exothermic effect at that high temperature. In isothermal tests or in adiabatic tests using relatively long "wait" times, the induction time for this type of decomposition at process temperature can be determined, normally resulting in a much lower onset point as shown in Figure 2.7B.

Figure 2.8 illustrates typical autocatalytic decompositions at three temperatures, 80°C, 90°C, and 100°C by isothermal tests.

Example 4: A Stability/Runaway Hazard Assessment Report
In Table 2.3, test data evaluating stability/runaway hazards of a material are presented. This table contains the relevant test data and evaluations in accordance with Section 2.1.5 (Test Strategies) and represents a good model for summary reports of such key information.

A number of important parameters influence chemical reactivity hazards such as:

- temperature–pressure variation,
- temperature–time variation,
- inhibitor depletion,
- catalytic effects of contaminations,

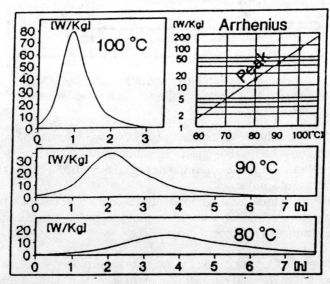

FIGURE 2.8. Typical Results of Autocatalytic Thermokinetics as Obtained by Isothermal Analysis.

2.1. SUMMARY/STRATEGY

TABLE 2.3 (Part I)
Example of Stability/Runaway Hazard Assessment Data and Evaluation Report

Test Substance Name:	4-Nitrophenol
Chemical Formula:	$C_6H_5NO_3$
Code:	X1Y2
Purity:	>99%

Literature Search and Theoretical Computations (Section 2.2):
1. Hazardous Groups: one nitro group, indicating potentially hazardous reactivity properties
2. Literature: compound determined to be nondetonable
3. Reactivity: incompatible with alkali materials; formation of very sensitive substances
4. Computer programs: CHETAH: $\Delta HH_d = -1.03$ kcal/g
                      Oxygen Balance: -132.27
                      Rating Criteria: 1: High (ΔH_d)
                                      2. Medium ($\Delta H_c - \Delta H_d$)
                                      3. Medium (Oxygen Balance)
                                      4. Medium (Y-criterion)
                        Energy Release Potential: **High**

Practical Testing (Section 2.3):
1. Screening Tests (Section 2.3.1)
        Type of Test: DSC
        Conditions: 10°C/min., high-pressure cups, both air and nitrogen tests
        Results   melting = start of melting approximately 95°C
                          $\Delta H_d = 2300$ J/g (549 cal/g)
                          $T_o = 275$°C

Evaluation (Section 2.1.5):
  Explosibility—detonation of pure substance not likely, deflagration cannot be excluded
  Thermal Stability/Runaway Potential—high energy compound, additional testing required
  Reactivity—self-reactive substance

2. Thermal Stability/Runaway Testing (Section 2.3.2)

Type of Test I:	Dewar test
Conditions:	Sample mass, 100 g
Results:	Adiabatic Induction Times: 2000 min at 180°C
	600 min at 200°C
	310 min at 210°C
	125 min at 225°C
	55 min at 240°C
Type of Test II:	ARC
Conditions:	Sample mass, 3 g $\Phi = 3.5$
Results:	ΔH_d: 500 J/g (120 cal/g) (lower limit because ARC runs were terminated before maximum temperature was achieved)
Results:	T_o: 240°C
Type of Test III:	Autoclave test (200 cm³ capacity)
Conditions:	$dT/dt = 1.25$ °C/min; degree of filling: A = 0.1 g/cm³ and B = 0.2 g/cm³
Results:	Runaway T_o A = 245°C B = 250°C
	$(dp/dt)_{max}$ A = 1400 bar/s B = 3000 bar/s
	p max A = 320 bar B = 773 bar

> **TABLE 2.3 (Part II)**
> **Example of Stability/Runaway Hazard Assessment Data and Evaluation Report**
>
> **EVALUATION (SECTION 2.3.2)**
>
> Based on the theoretical evaluation and practical testing, it is concluded that 4-nitrophenol is a thermally unstable compound. Based on its heat of decomposition, it cannot be excluded that the substance has deflagration properties. Based on literature evaluations, it has no detonation properties.
>
> Based on the DSC, ARC, and especially the Dewar test data, it is concluded that the substance can be handled safely in stirred systems with circulating cooling if the temperature of the substance does not exceed 160°C. Based on the kinetic data from the Dewar testing (E_a = 117 kJ/mol, $F(Q)$ = 7.6 × 10^{13} W/kg), the heat production (q) at that temperature is 0.59 W/kg. Alarm temperatures must be set at 170°C (q = 1.22 W/kg) and the cooling capacity of the system (per kg of material) must be able to cope with several times this heat production (see Chapter 3).
>
> From the autoclave testing, it is concluded that in case of a runaway, the substance produces a significant quantity of gases. The end stage of the runaway will certainly result in a thermal explosion. The venting evaluation should be conducted by the test methods and calculation procedures described in Chapter 3.

- catalytic effects of container material,
- increased volume of system,
- stronger initiation energy than used in tests, and
- increased surface contact among the reactants.

Different conditions exist in the laboratory, in the plant, in transportation, in field applications, and in disposal. These factors may have significant effects on the reactive properties of any chemical substance.

2.2 TECHNICAL SECTION

2.2.1 Thermodynamics

The stable equilibrium thermodynamic state of a system at constant pressure and temperature is the one with the minimum Gibbs free energy, G. This thermodynamic condition is defined as:

$$G = U - TS + pV \qquad (2\text{-}1)$$

where U is the internal energy of the system, T is the absolute temperature, S is the entropy, p is the pressure, and V is the volume. However, it should not be concluded that metastable (i.e., higher Gibbs free energy) thermodynamic states are not of practical importance despite the fact that such a system has a tendency to evolve spontaneously to its minimum Gibbs free energy state. In fact, an energy barrier, the so-called activation energy, E_a, may reduce signifi-

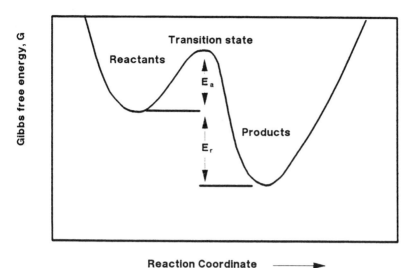

FIGURE 2.9. Schematic Energy Diagram of the Transition State Leading to Chemical Reaction

cantly the spontaneous evolution of the system to the minimum Gibbs free energy state. This is illustrated in Figure 2.9. In the present context, spontaneous exothermic decomposition or reaction may be negligible at standard conditions of 20°C and 1 bar due to an activation energy barrier.

Once an exothermic decomposition is initiated, usually by application of heat to raise the temperature, the energy that is released may maintain the higher temperature and thus cause the reaction to continue until all material is converted or until the reaction is stopped by forced cooling. The change in the Gibbs free energy during such a process (at constant temperature and pressure) is:

$$\Delta G_r = \Delta U_r - T\Delta S_r + p\Delta V_r \qquad (2\text{-}2)$$

Since the entropy change of an exothermic reaction is usually only a few tenths of a kilojoule per mol per °C [13], the factor $T\Delta S_r$ can be neglected particularly when the magnitude of ΔU_r is large. The energy change (ΔU_r) is equal to the difference between the energies of formation (ΔU_f) of the reactants and products:

$$\Delta U_r = \sum_{\text{products}} (U_f)_{\text{products}} - \sum_{\text{reactants}} (U_f)_{\text{reactants}} \qquad (2\text{-}3)$$

At constant temperature and pressure, ΔU_r is the observed energy dissipated or absorbed by the reaction. A negative value of ΔU_r means that energy is dissipated by the reaction. If the reaction occurs under isobaric conditions

TABLE 2.4
Structure of High Energy Release Compounds

Compound	Structure	ΔH_f° kJ/mol	ΔH_f° kJ/G
cyanogen	N≡C—C≡N	+308	+5.9
benzotriazole	$C_6H_4NH=N$ (ring)	+250	+2.1
nitrogen trichloride	NCl_3	+230	+1.9
acetylene	HC≡CH	+227	+8.7
allene	$H_2C=C=CH_2$	+192	+4.8
diazomethane	$H_2C=N^+=N^-$	+192	+4.6
hydrogen cyanide	HC≡N	+130	+4.8
1,3-butadiene	$H_2C=CHCH=CH_2$	+112	+2.1

Note: the numbers in the fourth column give a better relative indication of potential energy release than those in the third column.

(constant pressure), energy can then be replaced by enthalpy (ΔH_r). The relation between ΔU_r and ΔH_r is:

$$\Delta H_r = \Delta U_r + p\Delta V \tag{2-4}$$

Under isobaric conditions, a small amount of energy is consumed or released by expansion or contraction of the system, and ΔH_r is the observed (i.e., net) heat dissipated or absorbed.

The enthalpy released or absorbed in an isobaric process can be described in a manner similar to Equation (2-3) for constant volume conditions:

$$\Delta H_r = \sum_{products}(H_f)_{products} - \sum_{reactants}(H_f)_{reactants} \tag{2-5}$$

2.2.2 Identification of High Energy Substances

Substances with a positive enthalpy of formation will always release energy during their decomposition. Some typical structural similarities in high energy substances are the relative degree of unsaturation, high proportion or high local concentration of nitrogen in the molecular structure, and nitrogen-to-halogen bonds [31]. Examples are cyanogen, benzotriazole, acetylene, allene,

2.2 TECHNICAL SECTION

diazomethane, hydrogen cyanide, butadiene, and nitrogen trichloride. Properties of these substances are shown n Table 2.4.

Energetic substances can in general be identified by the presence of hazardous molecular structures [12, 31–34]. Peroxide groups, nitro groups, azo groups, double and triple bonds, and ring deformation and steric hindrance all influence the stability of a molecule. Compilations of energetic groups have been published by Bretherick [35] and by Leleu [36]. Table 2.5 lists a number of groups that have relatively weak bonds and that release substantial energy upon cleavage. The list on Table 2.5 is not exhaustive. The listing does, however, provide assistance in screening chemical structures for determining which should be investigated further before consideration of handling, even in small quantities.

The presence of one of the mentioned groups in the molecule does not necessarily imply that the substance is hazardous. For instance, a molecule that contains a nitro group attached to a long aliphatic chain does not show significant explosive properties. On the other hand, trinitromethane, which consists of three nitro groups attached to a methane group, does have dangerous explosive properties. "Diluting" the active groups by increasing the molecular weight decreases the explosive potential. A similar effect is obtained by mixing (diluting) active substances with inert materials. This effect is accounted for by ranking based on the decomposition energy per unit weight [10, 26], and by estimations with normalized data from group contributions [34].

However, the initial absence of unstable groups is no guarantee for long-term stability of the compound. For example, some aldehydes and ethers are easily converted to peroxides by reaction with oxygen from air [35, 37, 38]. Organic peroxides represent a class of unstable materials while monomers represent a class of substances that can self-react by polymerization if not properly inhibited and if the temperature is not properly maintained. Runaway reactions can result in both of these examples.

Example 1—Organic Peroxides
Organic peroxides are widely used as initiators and cross-linking agents for polymerization reactions. This group of materials characterized by the (–O–O–) bonding has shown a potential hazard varying from relatively low all the way to deflagration or detonation. The characteristic properties of this group [39] can be summarized as follows:

1. thermally unstable and sensitive to heat,
2. release of heat on decomposition,
3. sensitive to contamination,
4. formation of gases and sprays on decomposition,
5. formation of free radicals on decomposition,
6. limited oxidizing properties, and
7. deflagrative burning in absence of air.

TABLE 2.5
Typical High Energy Molecular Structures

Definition	Bond Grouping	Definition	Bond Grouping
acetylenic compounds	$-C \equiv C-$	tetrazoles; high nitrogen-containing compounds	$-N=N-N=N-$
haloacetylene derivatives	$-C \equiv C-X$	triazines (R=H, –CN, –OH, –NO)	$-C-N=N-N-C-$ $\quad\quad\quad\quad\quad\quad R$
metal acetylides	$-C \equiv C-M$	alkyl hydroperoxides; peroxyacids	$-C-O-O-H$
azo compounds	$-C-N=N-C-$	peroxides (cyclic, diacyl, dialkyl); peroxyesters	$-C-O-O-C-$
diazo compounds	$-C=N^+=N^-$	metal peroxides; peroxoacid salts	$-O-O-M$
diazeno compounds	$-C-N=NH$	amine chromium peroxocomplexes	$N \rightarrow Cr-O_2$
nitroso compounds	$-C-N=O$	azides	$-N_3$
nitroalkanes	$-C-NO_2$	halogen azides; N–halogen compounds; N–haloimides	$-N-X$
polynitro alkyl compounds; polynitro aryl compounds	$-C-NO_2$ $\quad\ \ \lfloor NO_2$	diazonium sulfides and derivatives; "Xanthates"	$-C-N=N^+\ S^-$
acyl or alkyl nitrates	$-C-O-NO_2$	diazonium carboxylates and salts	$-C-N=N^+\ Z^-$
acyl or alkyl nitrites	$-C-O-N=O$	amine metal oxo salts	$(N \rightarrow M)^+\ Z^-$
1,2–epoxides	$-C-C-$ $\ \ \lfloor O \rfloor$	N–metal derivatives	$-N-M$
metal fulminates	$-C=N-O-M$	halo-aryl metal compounds	$Ar-M-X$
aci-nitro salts	$HO-(O=)N=$	hydroxyammonium salts	$-N+-OH\ Z^-$
N-nitroso compounds	$-N-N=O$	arenediazoates	$-C-N=N-O-C-$
N-nitro compounds	$-N-NO_2$	arenediazo aryl sulfides	$-C-N=N-S-C-$
fluoro dinitromethyl compounds	$F-C-NO_2$ $\quad\quad\ \ NO_2$	*bis*-arenediazo oxides	$-C-N=N-O-N=N-C-$
difluoro amino compounds; N,N,N-trifluoroalkylimidines	$-N-F_2$	*bis*-arenediazo sulfides	$-C-N=N-S-N=N-C-$
N-azolium nitroimidates	$-N^+-N^--NO_2$		

Peroxides decompose rapidly with generation of heat [40]. This results in product loss and in rapid increases of temperature and pressure. Free radicals are formed during the decomposition, which induces further decomposition (autocatalytic decomposition). Insufficient heat removal results in a runaway, which may eventually be followed by an explosion and/or autoignition. The pressure has a marked accelerating effect on the decomposition process (deflagration rate). The thermal stability of peroxides can be expressed by the Self Accelerating Decomposition Temperature (SADT) values [41]. The decomposition energy is related to the active oxygen content. However, the mechanism of decomposition strongly depends on the molecular structure of the peroxide [39]. Organic peroxide decompositions are significantly catalyzed by small amounts of contaminants such as redox agents, strong ionizing agents, strong oxidizing and reducing agents, and heavy metals [28].

Example 2—Polymerization of Monomers
Polymerization is generally a highly exothermic process that can get out of control leading to a thermal explosion. The enthalpy of polymerization is usually about −20 kcal/mol (e.g., −17 kcal/mol for styrene). The reaction rate depends, in addition to other factors, on the mechanism of polymerization (e.g., free radical, ionic), the type of polymerization (e.g., vinyl, ring-opening), the concentration of monomer, the catalyst, and the temperature. Polymerization does not require the presence of catalysts (e.g., free-radicals), although they significantly increase the polymerization rate. Quite often, a combination of processes can occur. For example, in the case of styrene monomer, a high rate of polymerization in an isolated container can result in complete evaporation of the remaining monomer. Upon release of the vapor in air, a composition in the explosive range (1 to 6 volume % of styrene in air) may be formed [42].

If a container with styrene monomer is subjected to a large heat flux, for example, fire or steam, the polymerization of the monomer causes the temperature to rise. At a certain elevated temperature, spontaneous decomposition of the styrene monomer and/or its polymer starts. This secondary decomposition process generates twice as much energy as the polymerization process itself.

2.2.3. Hazard Prediction by Thermodynamic Calculations

In addition to reviewing lists of high energy materials such as presented in Section 2.2.2, the hazards of materials should also be estimated through various calculations including the use of computer programs. The important factors estimated are the oxygen balance and the decomposition energy. A general disadvantage of such calculations is the inability to predict kinetics.

2.2.3.1 Oxygen Balance
Vigorous oxidations are frequently explosive reactions. The reacting oxygen may be supplied from the atmosphere (e.g., in gas phase explosions) or from

liquid or solid oxidizers mixed with the substance, or may be available within the molecule itself. One estimating parameter for potential hazard and instability of substances or reaction mixtures containing oxygen is the oxygen balance [10, 43]. The oxygen balance is the amount of oxygen, expressed as weight percent, liberated as a result of complete conversion of the material to CO_2, H_2O, SO_2, Al_2O_3, N_2, and other relatively simple oxidized molecules.

A positive oxygen balance will be calculated if there is more than sufficient oxygen present in the molecule for complete conversion to the simple molecules. The oxygen balance will be zero if there is just enough oxygen present in the molecule to completely oxidize all of the reducing atoms in the molecule. If the amount of oxygen bound in the molecule is insufficient for a complete oxidation reaction, a negative oxygen balance results. The deficient amount of oxygen needed for the reaction to go to completion is reported with a negative sign.

For substances containing only carbon, hydrogen, and oxygen, when the oxygen balance is zero, the reaction is simply:

$$C_xH_yO_z \rightarrow xCO_2 + (y/2)H_2O$$

More generally, the oxygen balance (OB) of a substance $C_xH_yO_zN_q$ is expressed by Equation (2-6), when the nitrogen is assumed to evolve as N_2:

$$OB = \frac{-1600[2x + (y/2) - z]}{(MW)} \tag{2-6}$$

This equation is not valid for substances containing other elements.

Dinitromethane ($CH_2N_2O_4$), for example, has an OB by Equation (2-6) of +15.1. Acetic acid ($C_2H_4O_2$) has an OB of –106.7. This means that there is an excess of oxygen over that needed for the complete combustion of dinitromethane, whereas oxygen must be supplied in the case of acetic acid (i.e., 106.7 g of oxygen for the combustion of 100 g of acetic acid). The evaluation of the hazard potential based on the oxygen balance is not an absolute rating, but only an indicator. The significance of the rating depends on how the oxygen is bonded in the molecule. Acetic acid is certainly not an explosive material. The example shows that despite the presence of oxygen in the molecule, it could not be efficiently used for the oxidation of the carbon.

It is known empirically that explosives with an oxygen balance close to zero are the most powerful [44]. This statement also holds for mixtures of reducing and oxidizing substances for which the OB is formulated as:

$$OB = \frac{\sum_r N_r(MW)_r(OB)_r + \sum_o N_o(MW)_o(OB)_o}{\sum_r N_r(MW)_r + \sum_o N_o(MW)_o} \tag{2-7}$$

2.2 TECHNICAL SECTION

in which N is expressed in mols, subscript "r" refers to the reducing substance, and subscript "o" refers to the oxidizing substance.

The OB is used in a number of relationships with thermodynamic decomposition parameters. Yoshida [10] showed that there is a linear relationship between the absolute value of $|\Delta H_c - \Delta H_d|$ and the oxygen balance of a substance, where subscript "c" refers to combustion and subscript "d" refers to decomposition. Lothrop and Handrick [45] found a good correlation between the oxygen balance and the results from the lead block explosion test. Jain [46] made a distinction between oxidizing and reducing atoms in a molecule in analyzing oxygen balance relationships. He showed that for a stoichiometric reaction between various solid oxidizers and a number of organic fuels, a linear relationship exists between the heat of reaction and the number of oxidizing valences. Further, a good correlation appeared to exist between the number of reducing valences of organic fuels and the heat of combustion.

2.2.3.2 Calculation of the Reaction Enthalpy

The energy that is involved in a reaction can be estimated by calculation if the enthalpies of formation of the reactants and of the products are known. The enthalpy of formation, ΔH_f, is defined as the heat that is released during the formation of a substance from its elements at 20°C (or 25°C) and 1 bar.

The enthalpy of formation can be determined theoretically and experimentally. The theoretical methods can be defined as those which use bond contributions and the ones which use group contributions. The bond contribution techniques can be characterized as zero, first, second, or higher order methods, where zero is elemental composition only, first adds the type of bonding, second adds the next bonded element, and higher adds the next type of bond. A survey of typical theoretical methods is shown in Table 2.6.

A brief description of the calculation methods follows:

1. *Calculation of enthalpies of formation from the enthalpies of combustion of reactants and products:* This method is generally applicable for any combustible material for which the gross molecular formula is known. The enthalpies of combustion may be determined in a calorimeter using excess oxygen. Analysis of the combustion products may be appropriate.
2. *Method of Craven [50]:* This average bond energy summation method (ABES) is a simplification of the method described by Sanderson [55]. The reaction enthalpy is calculated by subtracting the total bond energies present before the reaction from the total bond energies of the products.
3. *Compilation of strength of chemical bonds by Kerr, Parsonage, and Trotman-Dickensen [47]:* This method is similar to the ABES method of Craven.

TABLE 2.6
Some Available Sources of Enthalpy of Formation Data
• Sources in literature [47–49]
• DIPPR[a] (source of thermal data)
• Determination of heat of combustion and analysis of combustion products [45]

• Bond contributions:
    1. Average bond energy summation method of Craven [50]
    2. Compilation of strength of chemical bonds by Kerr, Parsonage, and Trotman-Dickerson [47]

• Group contributions:
    1. Benson [49]
    2. Andersen, Beyer, Watson [51, 52]

• Computer programs: 1. CHETAH [11, 25–27] (see Section 2.2.3.3)
[a]DIPPR = Design Institute for Physical Property Data (AIChE) [b]CHETAH = Chemical Thermodynamic and Heat (Energy) Release Program (ASTM)

The available lists of bond types and equivalent energies are not limited to organic molecules. The calculation method, however, is essentially equal to the Craven method.

4. *Group contribution method of Benson [49]:* This method is used essentially in the CHETAH program. The molecule is subdivided into small groups represented by $A - (B)_i (C)_j (D)_k (E)_l$, where A is a particular atom or atomic group and B, C, D, and E are atoms or atomic groups attached to A with known enthalpies of formation. Benson gives corrections for ring compounds and for steric hindrance which make the method quite accurate.

5. *Group contribution method of Andersen, Beyer, and Watson [51, 52]:* In this method, a given compound is constructed from a base group (methane, cyclopentane, benzene, naphthalene, methylamine, dimethylamine, trimethylamine, or formamide) with known enthalpies of formation, which is then modified by appropriate substitutions to yield the desired molecule.

Results of calculations of enthalpies of formation using these methods, except for the enthalpies of combustion approach, are shown in Table 2.7.

From Table 2.7, it is concluded that the method of Kerr, Parsonage, and Trotman-Dickerson [47] shows large deviations in the calculated enthalpies of formation and is reliable only in the case of simple linear aliphatic molecules. The method of Craven [50] shows better results and has a wider scope of use. The largest errors occur when steric hindrance (TNT) or ring stress (cyclopropane) are involved. Methods 3 and 4, which are based on group contributions, show good results, both in the scope of use and the accuracy in

TABLE 2.7
Enthalpies of Formation (in kcal/mol) of 10 Chemicals Calculated by Five Methods at Standard Conditions of 20°C and 1 Bar

	Method					
Substance	Craven #1	K, P, & T #2	Benson* #3	A, B & W #4	CHETAH* #5	Exper. (gas)
acetonitrile	—[a]	—[a]	22.7	18.6	22.7	21.0
dichloroacetic acid	–118.8	–246.6	–111.0	–113.9	–111.0	–106.7
nitroethane	– 28.4	—[b]	– 25.2	– 23.9	– 24.5	– 24.2
ethylenediamine	– 7.8	5.5	– 3.6	– 5.3	– 3.6	4.2
cyclopropane	– 15.3[c]	– 41.7[c]	12.7	—[d]	12.7	12.7
pyrimidine	—[e]	—[e]	—[f]	—[f]	—[f]	47.0
1, 3-butadiene	33.4	– 70.2	29.7	30.1	26.3	26.3
2,4,6-trinitrotoluene	– 16.0	—[b]	– 17.8	– 6.6	9.8	2.1
t-butyl hydroperoxide	– 67.3	–147.6	– 57.6	—[g]	– 57.6	– 81.5
t-butyl methyl sulfide	– 29.7	47.1	– 29.5	– 28.9	– 29.5	– 28.9
Average unit deviation[h]	12.1	72.4	10.4	4.5	4.3	—
Number of compounds	8	6	9	7	9	—

*Temperature = 25°C; pressure = 1 bar
[a] –C≡N bond not included
[b] –N=O bond not included
[c] No ring deformation correction
[d] No cyclopropane ring in method
[e] –C=N– bond not included
[f] Some nitrogen groups not in table
[g] No peroxide group in table
[h] Average deviation from experimental values

predicting the enthalpies of formation. Benson [49] compensates for ring stress but does not take into account steric hindrance. In the method of Andersen, et al [51, 52], steric hindrance is accounted for but ring stress is not.

A comparative evaluation of various methods for predicting thermodynamic properties and enthalpies of combustion was made under the DIPPR program [56]. The CHETAH program was deemed, in general, the most accurate program and also the most broadly applicable. It does not generally compensate for ring stress and steric hindrance. The results shown in Table 2.7 confirm its applicability to estimating enthalpies of formation. Interestingly, none of the methods can really handle a complex heterocyclic such as pyrimidine because of the aromatic (double) bond between nitrogen and carbon.

Once the enthalpies of formation of the reactants and the products are known, the enthalpy of reaction or decomposition can be calculated according to Equation (2-5). Most of the decomposition products are small molecules such as H_2O, CO_2, CO, C, N_2, NO, NO_2, SO_2, and CH_4. However, the CHETAH program provides for user specification of the end products. Water can be formed as liquid or gas depending on the pressure and decomposition temperature. Carbon dioxide will be formed only if there is sufficient oxygen present in the molecule (i.e., OB ≥ 0), otherwise CO, CH_4, or carbon will be formed. The relative quantities of the reaction products are also determined by the decomposition temperature and pressure. The equilibrium reaction between CO and CO_2, for example, will shift to the left at increasing temperature, at decreasing oxygen, or at decreasing pressure:

$$CO + \tfrac{1}{2}O_2 = CO_2 + \text{heat}$$

Similar arguments can be applied for the formation of N_2 and its oxides. Enthalpies of decomposition, which are estimated on the basis of the products described above, usually result in a conservative prediction. In practice, decomposition is nearly always incomplete due to the evaporation of volatile reaction products and polymerization or tar formation by heavier molecules.

Consider, for example, the decomposition of *t*-butylperoxybenzoate (TBPB). Based on the set of small molecules described above, it could be assumed that methane, carbon, and water would be the main reaction products. However, a variety of decomposition products, including a significant amount of tar, have been determined experimentally, as shown in Table 2.8.

Substances with an OB of zero or more show a greater tendency to decompose into small molecules; thus, the theoretical and experimental values of enthalpies of decomposition will be in greater agreement. However, if the

TABLE 2.8
Decomposition Products of *t*-Butylperoxybenzoate (TBPB)

Substance	mol/mol TBPB	Substance	mol/mol TBPB
CO_2	0.9	biphenyl	0.1
propanone	0.5	phenyls/biphenyls	0.1
methane	0.2	ethyl benzene	0.05
toluene	0.2	1-phenyl-2-propanone	0.05
benzene	0.2	2-methyl-2-propanol	0.05
methane	0.1	tarlike residues	0.3
butanone	0.1		

2.2 TECHNICAL SECTION

OB of a substance is less than zero (a negative value), as is the case with TBPB, the calculated enthalpy of decomposition will generally be larger than the experimental value. A literature correction for the calculated value is the following [10]:

$$(\Delta H_d)_{corrected} = (\Delta H_d)_{calc} (0.0044 \text{ OB} + 0.96) \quad (2\text{-}8)$$

Other sources show the coefficient of 0.0044 in Equation (2-8) to range as wide as 0.0032 to 0.0095.

Apart from this point, the formula does lead to inconsistency with the enthalpy estimated by CHETAH, which calculates the maximum energy that is generated by decomposing a chemical and reacting the resulting elements to a combination with an optimum Gibbs free energy. In practice, detonations usually do not result in such an optimum energy release. In particular, detonable chemicals with a negative oxygen balance show lower decomposition energies. The formula as stated corrects for this phenomenon.

2.2.3.3 Application of Computer Programs

There are several computer software programs available for predicting the type and amount of reaction products, and for estimating the energy that is generated by decomposition and oxidation reactions [57]. These include CHETAH, TIGER, NASA, and REITP2, which are discussed in this section. Other available programs include NOTS-CRUISE, THEDIC [53] and TGAP [54], which are not discussed in this book. Only in the case of very high efficiency explosives do the products defined by CHETAH approximate the experimental values. Computations via CHETAH indicate the maximum energy yield as constrained by stoichiometry alone, which is used as a parameter for predicting impact sensitivity.

Except for CHETAH, which can estimate ΔH_f if required, these programs need information about the enthalpy of formation of the substance and the reaction products. This information must be input by the user or can be present in a data base. The programs are run mostly on mainframe computers, although CHETAH is also available in a PC version.

In general, none of these programs takes into account the complete molecular structure of the reacting molecules. The so-called zero order additivity method uses atomic contributions solely, in which the result is not influenced by the precise structure of the molecule. The first order method is commonly described as the bond contribution method, in which the molecular structure is considered to some extent. The second order method uses contributions of groups. The principle in prediction of the reaction products is based on the complete disintegration of the reaction molecules into atoms, and subsequent rearrangement of the atoms into small molecules. In the case of decomposition of TBPB illustrated previously in Table 2.8, the formation of toluene is, for example, not predicted by the computer programs. Instead, CH_4 and solid C are assumed as the reaction products. With some programs, for example, CHETAH, the products can be specified by the user.

The program REITP2 (Revised Evaluation of Incompatibility from Thermochemical Properties, Version 2) applies a list of reaction products in the order in which they are produced during the decomposition, independent of reaction temperature and pressure.

The TIGER and NASA programs evaluate the composition of the reactant mixture by minimizing the Gibbs free energy of the system at a given temperature and pressure. These programs contain data bases of possible reaction products with information about the respective enthalpies of formation, the respective entropies, and the temperature dependence of the specific heats. Further, the conditions under which a reaction occurs (isobaric, adiabatic, and so forth) can be changed.

In CHETAH, a linear programming technique is used to select the reaction products that give the maximum energy of decomposition. As stated previously, the user can select some or all of the reaction products based on literature, experiments, or intuitive reasoning. The temperature can be chosen by the user, but the pressure cannot be varied. The CHETAH program is based on the Benson method [49] to establish the enthalpy of formation of ideal gases. Included in CHETAH are the method of molecular group equations (group substitutions), and a large critically-assessed whole molecule data base.

None of the programs can predict kinetics, that is, the rate of reaction, the activation energy, or the order of the reaction. These parameters can only be determined experimentally. Except for CHETAH, the primary use of the programs is to compute the enthalpies of decomposition and combustion. In fact, acid–base neutralization, exothermic dilution, partial oxidation, nitration, halogenation, and other synthesis reactions are not included in the programs except for CHETAH, which can be used to calculate the thermodynamics of essentially any reaction.

A comparison of four calculation methods is shown on Table 2.9. Some details about each of the four computer programs follow.

CHETAH—Chemical Thermodynamic and Heat (Energy) Release Program
The CHETAH program, issued by the American Society for Testing and Materials, first appeared in 1974 [11, 25–27] to screen organic and organometallic chemicals and mixtures for their potential to undergo a violent decomposition leading to a deflagration or detonation. While it is still extremely useful for this application, industrial experience has shown that it is also a valuable tool for the estimation of thermodynamic data. The program estimates enthalpy, entropy, heat capacity, and Gibbs free energy data for specified chemical reactions. The next version to be released (Version 7) will also estimate data for a wide selection of inorganic salts. (*Note:* The availability of Version 7 was announced in 1994.) CHETAH classifies the energy hazard potential of a composition by a pattern recognition interpretation of the four internally generated criteria listed below:

TABLE 2.9
Comparison of Four Thermodynamic Calculation Computer Programs

	REITP2	NASA	TIGER	CHETAH
INPUT				
Substance	✔	✔	✔	✔
Molecular formula	—	—	—	✔
Functional groups	✔[a]	✔	✔	✔
Enthalpy of formation	—	—	✔	✔[a]
Options				
Combination[b] of $pT, pH, pS, pV, VT, VH, VS$	—	✔	✔	—
Choice of reaction products	—[c]	✔	✔	✔
Choice of isotherm	—	—[d]	✔	—
Displacement along isobaric/isochoric curves	—	—	✔	—
OUTPUT				
Reaction Products				
Molecular formula	✔	✔	✔	✔
Average molecular weight	—	✔	—	✔
C_p, C_v	—	✔	✔	✔
ΔH_r	✔	✔	✔	✔
ΔS_r or S	—	✔	✔	✔
ΔU_r	—	✔	✔	✔
Oxygen balance (OB)	✔	—	—	✔

✔ = included in computer program
— = not included in computer program
[a] in data file
[b] combination of constant conditions indicated
[c] restricted to either CO or CO_2
[d] ideal gas law equation only ($pV = nRT$)

1. *Chemical Energy Density* (maximum heat of decomposition constrained only by stoichiometry). For the first criterion, substances are placed in four different hazard classes based on enthalpy of reaction/decomposition as indicated in Table 2.10.

    In practice, ranks C and D in Table 2.10 do not recognize the fact that the chemicals under consideration still may have the potential to show rises in temperature of more than 600°C, which can result in significant and severe gas production and high pressures.

2. *Probability Correlation.* The second criterion is based on the difference between the enthalpy of combustion in excess oxygen and the maximum enthalpy of decomposition. This second criterion follows the

TABLE 2.10
Enthalpy of Decomposition or Reaction

Rank	Degree of Hazard	Maximum Enthalpy of Decomposition/Reaction[a] (kcal/g)	Possible Qualitative Interpretations of the Classifications
A	High	>0.7	Violently exothermic; detonation likely
B	Medium	0.3 to 0.7	Exothermic; detonation possible; deflagration likely
C	Low	0.1 to 0.3	Deflagration possible
D	Very low	<0.1	Propagation unlikely

[a]CHETAH uses the negative values of the decomposition/combustion enthalpies 1 = hexane

assumption that a substance with sufficient oxidizer within its own structure to convert fully to normal oxidation products presents a larger energy hazard potential than a substance that does not have sufficient oxygen. The hazard classifications based on a combination of criterion 1 and criterion 2 are shown in Figure 2.10. This was initially proposed in CHETAH [27] and later adopted by others.

In Figure 2.10, substances with a high hazard potential according to the first criterion are classified as such only if a difference between the enthalpy of combustion in excess oxygen and the maximum enthalpy of decomposition is less than 3 kcal/g. The hazard potential is changed to medium if the difference is between 3 and 5 kcal/g, and drops to a low hazard potential if the difference exceeds 5 kcal/g. Substances with a medium hazard potential according to the first criterion are classified as such only if the difference is less than 5 kcal/g. The hazard potential is changed to low if the difference exceeds 5 kcal/g.

Note that under criteria 1 and 2, CHETAH uses the negative values of the decomposition/reaction/combustion enthalpies, that is, energy release is indicated by a plus sign.

3. *The Oxygen Balance.* The oxygen balance related to oxygen available in the molecule, as defined and discussed in Section 2.2.3.1, is calculated using, for example, Equation (2-6) as the third criterion for hazard potential ranking in CHETAH. The classification criteria, according to CHETAH, are shown in Table 2.11, where a positive oxygen balance represents an excess net amount of oxygen available and a negative OB represents a shortage in the net amount of oxygen for complete oxidation. As discussed previously, it is known empirically that explosives with an oxygen balance close to zero are the most powerful [44].

2.2 TECHNICAL SECTION

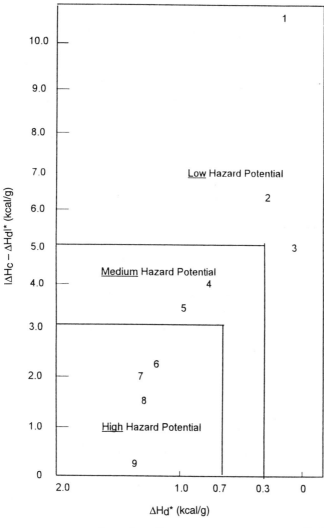

1 = hexane
2 = acetone
3 = acetic acid

4 = p-nitroaniline
5 = 2-amino-5-nitrophenylsulfone
6 = 2,4-dinitrophenol

7 = 2,4,6-trinitrotoluene
8 = picric acid
9 = nitroglycerine

*CHETAH uses the negative values of the decomposition/combustion enthalpies

FIGURE 2.10. Combination of Criteria 1 and 2 for Evaluating Explosibility in the CHETAH Program (including examples).

4. *The "Y" Criterion.* The fourth criterion for hazard potential classification in the CHETAH program takes into account the number of atoms in a molecule of the substance involved, and is called the "Y" criterion. It is defined as follows:

TABLE 2.11
Degree of Hazard in Relation to the Oxygen Balance
(CHETAH Criterion 3)

Hazard Potential	Value of OB
Low	OB < −240 or 160 < OB
Medium	−240 < OB < 0 or 80 < OB < 160
High	−120 < OB < 80

$$Y_{\text{criterion}} = \frac{10(Q^2)(MW)}{N} \qquad (2\text{-}9)$$

in which Q is the decomposition energy in kcal/g (with positive sign for heat released), MW is the molecular weight, and N represents the number of atoms in the molecule. The hazard classification is identified on Table 2.12.

REITP2—Revised Program for Evaluation of Incompatibility from Thermochemical Properties (version 2)
The computer program REITP2 has been described [10, 58]. The program was developed at Tokyo University to determine the potential hazard of a compound or a mixture of compounds (e.g., in storage) and is based on the OB of the substance(s), the enthalpy of reaction or decomposition, and the adiabatic temperature rise. In the case of a mixture of two or three substances, the program can calculate the most hazardous composition of the mixture. The calculation is based on Equation (2-5). The reaction products are obtained from a list that contains 300 different products in the order in which they are preferably formed. Reaction products with a high negative enthalpy of formation have a high position on this list.

The structural formula of the substance and its enthalpy of formation are input to the program, or the substance can be chosen from a data base

TABLE 2.12
Degree of Hazard in Relation to the Y-Factor
(CHETAH Criterion 4)

Hazard Potential	Value of Y-Factor
Low	Y < 30
Medium	30 < Y < 110
High	Y > 110

2.2 TECHNICAL SECTION

containing over 1100 substances with molecular composition and enthalpy of formation. Previously, it was shown that substances with a negative oxygen balance have an incomplete decomposition, so that a correction must be made for the calculated enthalpy of decomposition. If required, REITP2 can perform this correction on the basis of Equation (2-8). Options include selection of the formation of CO in preference to CO_2 and the formation of H_2O as a gas or liquid.

The REITP2 program produces the following output data:

1. OB of the substances,
2. amount and type of reaction products,
3. enthalpy of decomposition at standard conditions (25°C, 1 bar),
4. classification of substance into one of four ranks (see Table 2.10), and
5. OB and enthalpy of decomposition versus the composition of a mixture (only if two or three substances are involved).

The advantages and disadvantages of REITP2 are listed in Table 2.13.

TIGER Program

The TIGER program was developed at Stanford Research Institute [59] for calculating the thermodynamic state attained in a heterogeneous system of known atomic composition containing gases, liquids, and solids and in particular for detonation calculations. The program is primarily intended for use in the military and the explosives industry. The program can be applied for the determination of the thermodynamic state of nonideal heterogeneous systems in chemical equilibrium or partial equilibrium, and for detonation reactions. For the purposes of this Guidelines book, attention is focused

TABLE 2.13
Advantages and Disadvantages of REITP2

Advantages	Disadvantages
• The program comprises a large data base with over 1100 substances with enthalpies of formation	• Calculations are performed only for standard conditions (25°C, 1 bar)
• List of decomposition products contains 300 substances	• Predicted reaction products deviate from the experimental
• The program can quickly indicate the potential hazard of a substance or a mixture	• More limited number of elements than CHETAH
• Calculations of adiabatic temperature rise can be made	• Cannot estimate thermodynamic properties if required
	• User-specified reaction thermodynamics not feasible

primarily on the determination of the decomposition products and the enthalpy of decomposition under various conditions that can be calculated through the TIGER program.

The equilibrium state is generated by minimizing the Gibbs free energy of the system at a given temperature and pressure. In [57], the method is described as the modified equilibrium constant approach. The reaction products are obtained from a data base that contains information on the enthalpy of formation, the heat capacity, the specific enthalpy, the specific entropy, and the specific volume of substances. The desired gaseous equation of state can be chosen. The conditions of the decomposition reaction are chosen by defining the value of a pair of variables (e.g., p and T, V and T). The requirements for input are:

1. the formula of the substance,
2. the enthalpy of formation,
3. the molar volume,
4. the molar entropy, and
5. composition of the mixture (if necessary).

NASA-CET Program

The NASA-CET program was developed at the NASA Lewis Research Center [60]. The program has about the same features as the TIGER program, but incorporates higher temperatures such as are present in flames.

2.2.4 Instability/Incompatibility Factors

Two substances which have no hazardous reactivity properties in themselves can become dangerous when mixed. Certain groups of chemicals are likely to react with common substances such as air, water, acids, alkalies, and metals. Information about the possibility of such reactions is available in manuals on hazardous chemical reactions [35, 61, 62]. Examples of substances having incompatibility hazards when mixed are shown on Table 2.14. Applications of CHETAH to mixture instability determination [63, 64] and to binary incompatibility [65] have been published.

2.2.4.1 Factors Influencing Stability

Factors such as temperature, concentration, impurities, confinement, and the presence of air (oxygen) or solvents significantly influence the stability of a chemical substance relative to decomposition stability [31]. A discussion of these factors follows:

- *Temperature*—Temperature is the most important factor influencing reaction rate as shown in the Arrhenius equation, which describes the appropriate relationship indicating temperature as an exponential term

TABLE 2.14
Examples of Hazardous Incompatibility Combinations

Substance A	Substance B	Potential Phenomenon
Oxidizing agent	Combustible	Formation of explosive mixture
Chlorate	Acid	Hypergolic[a] ignition
Chlorate	Ammonium salt	Formation of explosive ammonium salts
Potassium chlorate	Red phosphorous	Formation of explosive mixture sensitive to shock and friction
Potassium chlorate	Sulfur	Formation of explosive mixture sensitive to shock and friction
Chlorite	Acid	Hypergolic[a] ignition
Hypochlorite	Acid	Hypergolic[a] ignition
Anhydrous chromic acid	Combustible	Hypergolic[a] ignition
Potassium permanganate	Combustible	Hypergolic[a] ignition
Potassium permanganate	Concentrated sulfuric acid	Explosion
Carbon tetrachloride	Alkali metal	Explosion
Nitro compound	Alkali	Formation of very sensitive substance
Nitroso compound	Alkali	Formation of very sensitive substance
Nitroso amine	Acid	Hypergolic[a] ignition
Alkali metal	Water	Hypergolic[a] ignition
Hydrogen peroxide (aqueous solution)	Amine	Explosion
Ether	Air (oxygen)	Formation of explosive organic peroxides
Olefin hydrocarbon	Air (oxygen)	Formation of explosive organic peroxides
Nitrite	Ammonium salt	Formation of explosive ammonium salts
Acetylene	Copper	Formation of copper acetylide sensitive to shock and friction
Picric acid	Lead	Formation of lead salt sensitive to shock and friction
Concentrated nitric acid	Amine	Hypergolic[a] ignition
Sodium peroxide	Combustible	Hypergolic[a] ignition

[a] Hypergolic: spontaneous ignition upon mixing.

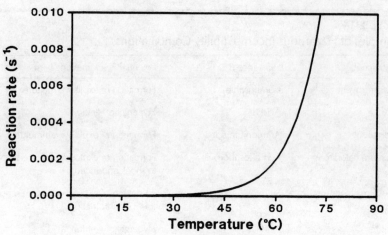

FIGURE 2.11. Reaction Rate as a Function of Temperature (Arrhenius Equation)

(see Figure 2.11). In practical terms, this means that an increase in temperature of 10°C will increase a specific reaction rate by two to four times depending on the energy of activation. At higher temperatures, side or consecutive reactions can be initiated with a different set of kinetic parameters.
- *concentration of reacting materials or decomposing material*—In general, the reaction rate is proportional to the concentration of the reactant(s). Dilution of an unstable substance with an inert solvent may be used to reduce the reactant concentration and even to minimize the temperature rise by the evaporation of the solvent. In addition, the adiabatic temperature rise is decreased. Accumulation of reactants (e.g., as a result of insufficient mixing), too rapid a charging rate, or too low a temperature for the reaction to start, may, on the other hand, lead to a higher adiabatic temperature rise. These aspects are treated more extensively in Chapter 3.
- *Impurities with catalytic effects*—Impurities that act as catalysts, reducing the activation energy of a process, may increase the rate of reaction significantly, even when present in small quantities. The presence of sulfuric acid, for example, increases the rate of decomposition and decreases the observed onset temperature of various isomers of nitrobenzoic acid [28]. Also, other substances such as NaCl, $FeCl_3$, platinum, vanadium chloride, and molybdenum chloride show catalytic effects. As a result, the decomposition temperature can be lowered as much as 100°C. Catalysts, such as rust, may also be present inadvertently. Some decomposition reactions are autocatalyzed, which means that one of more of the decomposition products will accelerate the decomposition rate of the original substance.

- *Solvents*—In addition to the diluting aspect, relatively low-boiling solvents can control a reaction temperature. The solvent will start to boil and can reflux through a condenser system if the reaction temperature reaches its boiling point. The reflux process prevents a further increase of the reaction rate by effectively limiting the temperature to approximately the boiling point of the solvent. Such a process also serves to remove the evolved heat from the reaction as the enthalpy of vaporization of the solvent. This process of controlling the reaction temperature is discussed further in Chapter 3. It should be noted that, in general, enthalpies of vaporization are much larger, on a mass basis, than are enthalpies of reaction for organic systems. Water should also be considered as a solvent since its enthalpy of vaporization (540 cal/g) is significantly higher than essentially all organic solvents. Care should be taken, however, to ensure that potentially hazardous substances (e.g., peroxides) do not accumulate after evaporation of some of the solvent (see Chapter 3),
- *Air*—As mentioned later in Section 2.2.4.2, many substances are capable of forming peroxides in contact with the oxygen in air. Inadvertent oxidation reactions cause quality loss, potential self-heating, and possibly ignition with a resultant fire or explosion. Also, oxygen can play a key role in the depletion of inhibitors in vinyl monomers resulting in uncontrolled polymerizations.
- *Confinement*—Deflagration rates of substances such as azo compounds, peroxides, and certain lead oxides may accelerate by pressure increase, especially when the governing decomposition reaction is gas-phase controlled [28]. Initiation of a deflagration at the bottom or at the center of a closed or partially closed vessel may lead to an increase of eh deflagration rate by a factor of more than 100 in comparison with top initiation. Autocatalytic decomposition by a volatile catalyst is enhanced by confinement.

2.2.4.2 Redox Systems

Five types of redox systems can be characterized as follows:

1. reactions between combustibles and oxygen from air,
2. redox compounds which are formed between oxidizing species and reducing species,
3. reactions between organic compounds and oxidizers other than oxygen,
4. reactions of strong reducing agents with organic compounds, and
5. reactions between strong reducing agents and strong oxidizers.

Following are comments on each of these five types.

1. Reactions between Combustibles and Oxygen from Air

By far the most important redox reaction relative to chemical stability is the reaction between an oxidizable material and oxygen from air. The particle size and any droplets have a large effect on the combustion properties. Some substances react so rapidly in air that ignition occurs spontaneously. These so called pyrophoric compounds (white phosphorus, alkali metals, metal hydrides, some metal catalysts, and fully alkylated metals and nonmetals) must be stored in the absence of air.

Some chemicals are susceptible to peroxide formation in the presence of air [10, 56]. Table 2.15 shows a list of structures that can form peroxides. The peroxide formation is normally a slow process. However, highly unstable peroxide products can be formed which can cause an explosion. Some of the chemicals whose structures are shown form explosive peroxides even without a significant concentration (e.g., isopropyl ether, divinyl acetylene, vinylidene chloride, potassium metal, sodium amide). Other substances form a hazardous peroxide on concentration, such as diethyl ether, tetrahydrofuran, and vinyl ethers, or on initiation of a polymerization (e.g., methyl acrylate and styrene) [66].

2. Redox Compounds Formed by Oxidizing and Reducing Species

Redox compounds that contain both reducing and oxidizing groups in their molecules are, for example, tin (II) perchlorate, peroxyformic acid, ammonium dichromate, and the double salt potassium cyanide with potassium nitrite [43].

TABLE 2.15
Structures Susceptible to Peroxidation in Presence of Air

Structure	As in
$-C-O-C-$	Ethers; cyclic ethers; acetals
$-CH=O$	Aldehydes
$CH_2=CH-CH_2-$	Allyl compounds
$-HC=CH-X$	Haloalkenes
$-HC=CH-$	Olefins
$-HC=CH-CH=CH-$	Dienes
$-HC=CH-C\equiv C-$	Vinyl acetylenes
$\Phi-CH=CH_2$	Styrene and similar types
$\Phi-CH-(CH_3)_2$	Cumene and similar types
$-C-NH-CO-NH_2\ N-$	Alkyl ureas
$-C-CO-NH-C-N-$	Alkyl amides; lactams

2.2 TECHNICAL SECTION

3. Reactions between Organic Compounds and Oxidizers Other Than Oxygen

Common industrial oxidants are nitric acid, sulfuric acid, chlorine, chlorates, perchlorates, iodates, periodates, bromates, hydrogen peroxide, peroxy acids, ozone, nitrates, nitrites, permanganates, persulfates, and dichromates. In [67], a number of oxidizing agents are classified according to their stability. Also presented are further storage and handling precautions together with some case histories showing the problems associated with the storage of these types of materials.

Yoshida [57] has evaluated a number of oxidants by calculation of the adiabatic temperature rise of a mixture of oxidizer and various organic compounds. Theoretically, pure oxygen provided the highest adiabatic temperature rise, followed by NO_2, $KClO_3$, HNO_3, NH_4ClO_4, NH_4NO_3, air, $K_2Cr_2O_7$, and $KMnO_4$. Nitric acid may oxidize materials at relatively low temperatures and concentrations. The nitrous gases evolved may lead to significant pressure development. Hydrogen peroxide may, even in low concentrations, , be catalytically decomposed by traces of corrosion products such as heavy metals. The liberated oxygen may lead to oxygen enrichment and pressurization of the reactor space. Hydrogen peroxide in combination with organic solvents may lead to the formation of organic peroxide compounds of an explosive nature. Chlorine has caused several incidents in industry as a result of reaction with organic substances such as polypropylene (used in filter elements), silicone oils, dibutyl phthalate (used in pressure transmitters), polychlorinated biphenyl heat transfer fluids (little used now), hydrocarbon oils (commonly used in pumps for chlorine service), glycerine, and waxes [68].

4. Reactions of Strong Reducing Agents with Organic Compounds

Alkali metals, finely divided aluminum and magnesium particles, hydrazine, diborane, metal hydrides, and hydrogen are strong reducing agents [35]. An example of a significant problem is the possible explosive reaction between light metals and carbon tetrachloride which is itself a stable compound [57].

5. Reactions between Strong Reducing Agents and Strong Oxidizers

Reactions involving strong oxidizers (pure oxygen, NO_2, HNO_3, and NH_4ClO_4) and strong reducing agents, such as propellant systems used for rockets, are exceptionally hazardous. These systems are able to generate extremely large energy releases.

2.2.4.3 Reactions with Water

A number of substances react vigorously with water, sometimes with the formation of hydrogen gas, which itself may ignite in the presence of air. Examples or such reactants are alkali metals, finely divided light metals and their hydrides, anhydrous metal oxides, anhydrous metal halides, nonmetal halides, and nonmetal oxides as well as certain organics such as anhydrides and isocyanates. Significant heat may be generated by dissolution of compounds such as potassium hydroxide, sulfuric acid, and hydrochloric acid in water.

2.2.4.4 Reactions between Halogenated Hydrocarbons and Metals

Halogenated hydrocarbons are generally regarded as relatively nonhazardous with only a modest level of potential energy. However, under certain conditions, violent reactions can occur between halogenated hydrocarbons and light metals, particularly aluminum and zinc [69]. The mechanism of the reaction is not exactly clear, but one mechanism considered is the reaction of the halogenated hydrocarbon with the metal to form a metal halogenide. This metal halogenide then acts as a catalyst for an exothermic polymerization reaction.

For example, the reaction between alumnium and tribromomethane may be described as follows:

$$Al + CHBr_3 \rightarrow Al\text{-alkyl bromide} + AlBr_3$$

$$CHBr_3 \xrightarrow{AlBr_3} \text{polymer} - \Delta H_r$$

A practical example is the reaction between t-butylchloride and magnesium turnings in ether for the preparation of a Grignard reagent:

$$Mg + (CH_3)_3CCl \xrightarrow[\text{(in ether)}]{} (CH_3)_3CMgCl$$

The heat generated is dissipated by reflux of the ether. The hazard of this reaction relates to the large energy release, the unpredictable nature of the reaction initiation, and the general problems of using ethyl ether.

2.3 PRACTICAL TESTING

2.3.1 Screening Tests

2.3.1.1 Thermal Analysis

Differential thermal analysis (DTA) is a technique in which the temperature difference between a substance and reference material is measured as a function of temperature or time while the substance and reference material are subjected to a controlled increase in temperature. Differential scanning calorimetry (DSC) is a technique in which the difference in energy inputs into the sample and reference material required to keep their temperatures equal is measured as a function of temperature while the substance and reference material are subjected to a controlled increase in temperature [70].

Classical DTA has been developed into heat-flux DSC by the application of multiple sensors (e.g., a Calvet-type arrangement) or with a controlled heat

2.3 PRACTICAL TESTING

leak (Boersma-type arrangement). The power-compensation DSC is another type in which the energy inputs to the reference and sample are adjusted to remove the temperature difference between the two materials. Schematic representations of a heat-flux DSC and a power-compensation DSC are shown in Figure 2.12.

The purpose of differential thermal systems is to record the difference in the enthalpy changes that occurs between the reference and the test sample when both are heated in an identical fashion. Several publications are available concerning the theoretical aspects and applications of various thermal analysis techniques, including the DSC [71–74]. Commercial instruments are available from a number of companies including Perkin-Elmer, TA Instruments, Toledo-Mettler, SETARAM, Seiko, and Polymer Laboratories.

In studying the hazard potential of substances, DSC is used to establish approximate temperature ranges in which a substance undergoes an exothermic decomposition or is susceptible to oxidation, and to determine the enthalpies of the appropriate reaction. The tests are performed with and without air to distinguish the thermal hazards due to oxidation and decomposition respectively. The oxidation rate is an indication of self-heating properties in the presence of air which is of importance, for example, in storage, drying, and sifting of powders. The decomposition without the presence of oxygen is meaningful for reviewing explosibility. There is seldom enough oxygen present in a sealed DTA or DSC cell to indicate the true potential for oxidation. If a significantly lower observed onset temperature is observed when testing in the presence of oxygen, further tests are probably required. In some publications [10, 75, 76], the results of DSC tests are correlated with the results of internationally accepted explosibility and thermal stability tests [77–79].

Another useful application of DSC is the determination of the specific heat (C_P) of substances as a function of temperature [71]. The specific heat is an important parameter in many thermodynamic and process design calculations.

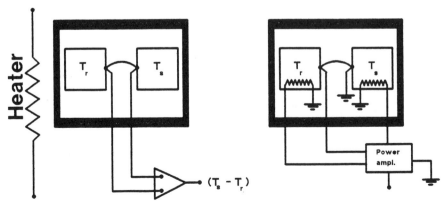

FIGURE 2.12. Schematic Representation of Heat-flux DTA and Power Compensation DSC

Most of the DSC equipment can be used in the temperature range of 25°C to 500°C. Most can be cooled as well, a feature required for investigating samples that are unstable at ambient conditions. DSC equipment is usually sufficient for indicating thermal hazards of stirred systems and small-scale unstirred systems provided the reaction is kinetically controlled under normal operating conditions, but the resulting data must be used with careful judgment if mixing or mass transport are important.

Performance of a DSC experiment
In DSC experiments, closed cells (pans) are used to prevent sample evaporation. These cells are sealed high pressure containers made of various materials and styles such as aluminum, stainless steel (SC-DSC), sealed glass-lined cups, gold plated cups (to avoid catalytic effects), or glass capillaries. A method that reduces, but does not eliminate evaporation, is the use of open cells under pressurized conditions (inert gas) in the high pressure DSC. In HP-DSC, a pressure of 10 bar has been found adequate for the purpose of general thermal stability screening tests [80]. DSC experiments aimed at detecting exothermic reactions involving solids with oxygen are conducted in open sample cups either static under high oxygen pressure or with a continuous flow of air. This technique is not used to quantify enthalpies of combustion.

The generalized DSC procedure involves placing a representative sample of usually less than 10 mg (which is not always easy to get!) in the sample cup. This cup is, if necessary, sealed and inserted in the DSC equipment on the sample sensor side. The reference cup, identical to the sample cup but containing the same amount of inert reference material (e.g., glass beads) with approximately the same C_p as the sample, is placed on the reference sensor side. In studying the temperature range of exothermic decomposition, a heating rate of 5 to 10°C/min (scanning mode) is generally used. Both power compensated DSC and heat-flux DSC are applicable in studying the hazard potential of substances; these methods yield similar results when testing for reaction/decomposition exotherms.

The observed onset temperature, T_o, is the temperature at which the substance or mixture first shows an observable instrumental response due to decomposition or reaction. The value of this T_o depends on the sensitivity of the apparatus, the sample mass, the atmosphere, the confinement, and the heating rate. When the experiment is designed to establish the onset temperature of the exotherm with more accuracy, a heating rate of 1 to 5°/min is appropriate [23, 77, 81, 82]. However, it should be emphasized again that the onset temperature value strongly depends on the instrument sensitivity and that application of onset temperature, and kinetic data, obtained in the DSC to large-scale operations may introduce significant errors.

An example of an idealized DSC curve with an exotherm peak is represented in Figure 2.13. The shape of the DSC curve depends on the reaction order, the occurrence of autocatalytic decomposition, and on parameters such

2.3 PRACTICAL TESTING

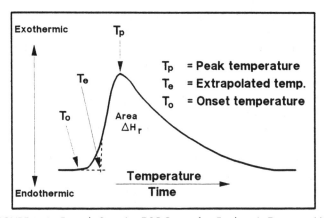

FIGURE 2.13. Example Scanning DSC Curve of an Exothermic Decomposition.

as heating rate, heat capacity, and the heat conductivity of both sample and reference materials. The peak area is proportional to the total enthalpy of decomposition or reaction.

Because of the possibility of unrecognized exothermic reactions at higher temperatures, DSC tests should be run at 400°C to determine if potentially catastrophic exotherms occur in a region that can be entered by having an uncontrolled exotherm in the lower temperature range.

DSC can be used effectively in the isothermal mode as well. In this case, the container with the sample is inserted into the DSC preheated to the desired test temperature. This type of experiment should be performed to examine systems for induction periods that occur with autocatalytic reactions and with inhibitor depletion reactions. (Reactions with induction periods can give misleading results in the DSC operated with increasing temperature scans.) Autocatalytic reactions are those whose rates are proportional to the concentration of one or more of the reaction products. Some hydroperoxides and peroxy esters exhibit autocatalytic decomposition. Inhibitor depletion can be a serious problem with certain vinyl monomers, such as styrene and acrylic acid, that can initiate polymerization at ambient temperatures and then self-heat into runaways. Isothermal DSC tests can be used to determine a time to runaway that is related to the inhibitor concentration.

For single reactions with uncomplicated kinetics and with availability of a truly representative sample, the DSC can be used with different scan speeds (temperature/time) to determine kinetic constants in the Arrhenius equation. This method, proposed by Ozawa [83] has been accepted by the ASTM Method E698. After determining kinetic constants by this method, it is desirable to check the constants by running an isothermal DSC aging test for a period of time followed by a DSC scan to see if the predicted fraction decomposition occurred in the aging step. Combinations of isothermal and scanning DSC experiments are required to examine in detail the kinetics of decomposition.

For reaction systems without solvents, the time at the test temperature is important. In these cases, the elected period of time and temperature should be based on the worst case operating conditions.

The melting transition of ultra-pure metals is usually used for calibration of DSC instruments. Metals such as indium, lead, and zinc are useful and cover the usual temperature range of interest. Calibration of DSC instruments can be extended to temperatures other than the melting points of the standard materials applied through the recording of specific heat capacity of a standard material (e.g., sapphire) over the temperature range of interest. Several procedures for the performance of a DSC experiment and the calibration of the equipment are available [84–86]. A typical sensitivity of DSC apparatus is approximately 1 to 20 W/kg [15, 87].

Assessment of the results: instability/onset temperature

The first item determined from a scanning DSC experiment is the observed onset temperature (T_O) of the exotherm. Since this observed onset temperature depends on the scanning rate and on the sensitivity of the instrument, its use is limited in establishing a safe operating region. Such a determination is usually done with an understanding of the operational and equipment specifications combined with the kinetics of the reaction or decomposition. As a rule-of-thumb in testing by DSC, s substance is indicated to be sufficiently thermally stable from decomposition (Box 10, Figure 2.3) if the T_O of the exotherm, as determined in the DSC, exceeds the highest process temperature by at least 100°C, recognizing other factors such as short time periods and the existence of good heat transfer conditions [10, 77, 81, 82, 88, 89]. However, exceptions to this rule-of-thumb have been shown [90].

As outlined in Section 2.2.4.1, impurities and catalysts may decrease T_O significantly; a decrease of 100°C is not unusual. The material of construction of the sample cup may act as a catalyst, resulting in surface-induced decomposition which may even be promoted by the sample/surface area ratio in the DSC cup. Therefore, it is important to check if the substance is catalyzed during the DSC experiment and if such catalysis is representative of process conditions. Frequently, substances that are sensitive to catalysis are handled in passivated glass-lined reactors, receptacles, or containers. Another phenomenon to recognize is autocatalytic decomposition. Substances that are susceptible to autocatalytic decomposition have an induction period prior to initiation of rapid decomposition. The same holds for substances that contain inhibitors, which can be depleted.

A DSC run in scanning mode does not provide the proper experimental conditions for autocatalytic decomposition to be identified as such because there is a continuous increase in temperature in the operation. In particular, the test does not determine the true thermal characteristics of autocatalytic

2.3 PRACTICAL TESTING

reactions, that is, the existence of an induction period and its temperature dependence. For autocatalytic substances held isothermally, decomposition will be observed following an induction period at temperatures well below an observed DSC onset temperature obtained in a scanning experiment. Autocatalytic decomposition can be determined with DSC tests in the isothermal mode using the isothermal step method [82].

Assessment of results: heat production/enthalpy of decomposition

In DSC instruments, heat production (q) can be determined directly as a function of temperature. The shape of the heat production curve is also important for hazard identification. A sharp rise in energy release rate (i.e., a steep slope of the exotherm), whether due to a rapid increase of the rate constant with temperature or to a large enthalpy of reaction, indicates that the substance or reaction mixture may be hazardous. Figure 2.14 illustrates an example of a DSC curve with a gradual exothermic reaction, while Figure 2.15 is an example of a steep exothermic rise.

The enthalpy of decomposition is determined by integrating the peak area (A_p) above the base line of the scan, using the following equation:

$$\Delta H_d = \frac{[K(T)][A_p]}{m} \qquad (2\text{-}10)$$

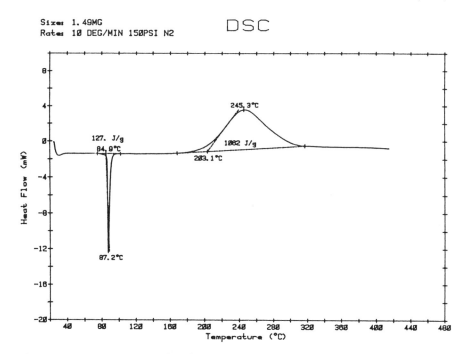

FIGURE 2.14. DSC Curve—Typical Exothermic Reaction

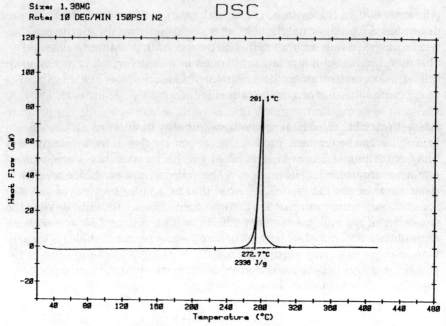

FIGURE 2.15. DSC Curve—Steep Exothermic Rise.

where

$K(T)$ = apparatus calibration constant, cal/m^2
A_p = peak area, m^2
m = sample mass, kg

Assessment of the results: kinetics
In principle, DSC offers the capability of calculating the kinetics of an exothermic reaction or decomposition from isothermal and scanning experiments.

Several calculation methods for establishing reaction kinetics from the scanning DSC results are discussed in the literature [91-97]. Kinetic constants can be obtained from scanning DSC tests under favorable circumstances that include: (1) the availability of a sample that even in milligram quantities is truly representative of the process material, and (2) kinetics that are uncomplicated by multiple reactions or by the presence of autocatalysis or inhibitor effects.

Calculation methods are either differential or integral. Differential methods use the DSC signal, which is directly proportional to the rate of reaction as a function of the temperature. Examples of such methods are Borchardt and Daniels [98], Kissinger [99, 100], and Ozawa [83]. Integral methods use the integrated DSC signal. This is directly proportional to the conversion as a function of the sample temperature. Examples of integral methods are Coats and Redfern [101] and Horowitz and Metzger [78].

2.3 PRACTICAL TESTING

Advantages/disadvantages of DTA/DSC

The main advantages of the DSC as a screening test are that it is a fast technique, requiring only a few hours for a complete scanning experiment, and that it requires only a few milligrams of sample, which is useful when only small quantities are available early in a research stage. The small sample size permits the use of DSC in a regular laboratory set-up even for very energetic systems. Furthermore, DSC is applicable over a wide temperature range. With DSC, the complete temperature range for thermal decomposition or reaction can be scanned, up to 400°C, or even higher in some instruments. Thus, enthalpies of decomposition or reaction can be established from a single scanning experiment. (However, a single scan cannot be recommended to determine kinetics.) These benefits make the DSC a favorite as a screening test.

Although small sample sizes are desirable from safety and availability standpoints, a small sample of a heterogeneous system may not be truly representative. Catalytic effects of the sample cup construction material may play an important role in the decomposition process. Determination of the enthalpy of reaction can be done by DSC if stability of the system permits. Autocatalytic effects cannot be accurately determined in a scanning experiment. It must be emphasized that the kinetic data acquired with a DSC experiment are only an indication of the behavior of the decomposition or global (overall) reaction occurring in the test cell. Without verification of the chemistry performed in the test cell, the kinetics can only be viewed as global (i.e., not for a specific chemical reaction mechanism). Obtaining kinetic data in the scanning mode is usually successful only for singular mechanistic systems. Kinetic data obtained for complex reactions cannot be readily measured, interpreted, or scaled.

Data on the enthalpy change, ΔH_d or ΔH_r, established in testing homogeneous samples can be used in the extrapolation to large scale. In the DSC experiments, no pressure data are obtained.

For full-scale processing, more accurate and more dedicated techniques using larger samples may be necessary, following the specific experimental hazard evaluation testing scheme outlined previously in Figure 2.4.

2.3.1.2 Isoperibolic Calorimetry

Constant jacket temperature measuring techniques, known as isoperibolic calorimetry, are designed to investigate the thermal behavior of substances and reaction mixtures under processing conditions [89, 102–108].

Isoperibolic equipment consists of a sample container that is placed in a circulating air oven or heater. Typically, the sample container, of which there are several types, consists of a small tube or beaker that can contain a 5 to 30 g sample. The temperature range of commercial instruments is about 0 to 300°C. Some instruments include small autoclaves or small stirred vessels. Open vessels are made of glass while autoclaves are constructed of stainless steel. Also available are a combination of a glass vessel, fitted with a glass vent, placed in a steel jacket, and glass tubes that are connected to a pressure transducer. The pressure range of test-tube autoclaves is about 1 to 100 bar. Some test apparatus can provide gas flow through the sample.

In an isoperibolic experiment, the jacket temperature of the sample container (or the surroundings of the container, i.e., the oven temperature) is held constant. On attaining a steady-state, a temperature difference between the sample and jacket may be obtained, which becomes: (1) zero (within the detection limit of the equipment) if no energy is released from the sample, or (2) positive if energy is released due to chemical reaction or decomposition. If no temperature difference is recorded after a fixed time interval, the oven temperature is increased (typically 5°C) and held constant once again. This procedure is repeated until an exothermic event is observed.

For the identification of the onset temperature of the exotherm, the steady-state temperature difference may be plotted against the sample temperature. After calibration, the evolved heat can be estimated. A typical plot of an isoperibolic measurement is illustrated in Figure 2.16. The sample is heated by step-wise adjustment of the jacket (or oven) temperature. The actual sample temperature results from the heat accumulation as net difference between the heat generated by the chemical reaction and the heat transferred to the jacket (or oven). The resulting mean temperature difference is relatively small and not easy to detect accurately. Thus, a range of step changes in temperature is used to define a curve, which enables a more accurate determination of the start of the exothermic event and of T_0 to be made.

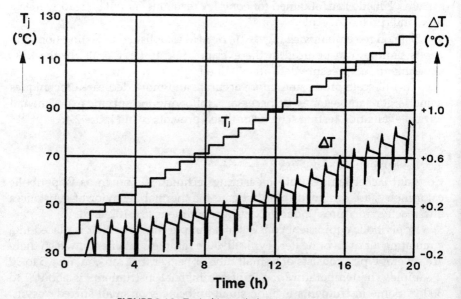

FIGURE 2.16. Typical Isoperibolic Measurement.

2.3 PRACTICAL TESTING

Commercially available instruments include the SEsitive Detector of EXothermic processes (SEDEX) [103, 104], SIKAREX [106], with a typical sensitivity of 0.5 W/kg), and RADEX [102, 108]. This equipment can also be run inisothermal and scanning modes.

Assessment of results
The onset temperature of exothermic activity, data on the autocatalytic decomposition, and the induction time for exothermic decomposition can be obtained. Also, it is possible to get pressure–time data with the correct instrument.

Because of the medium or relatively large sample quantities used and the instrumental sensitivity, isoperibolic calorimetry is a useful tool in determining the onset temperature of an exotherm. In fact, in its simplest construction, this is really the only measurement. Digital data acquisition does allow computer analysis of the peak or area under the curve, which indicates the order of magnitude of the exotherm. Generally, the detected onset temperatures are similar to those found in the ARC (see later in Section 2.3.2.3) and are significantly lower than in the DSC (Section 2.3.1.1) [79].

Isoperibolic instruments have been developed to estimate enthalpies of reaction and to obtain kinetic data for decomposition by using an isothermal, scanning, or quasi-adiabatic mode with compensation for thermal inertia of the sample vessel. The principles of these measuring techniques are discussed in other sections.

Advantages/disadvantages: isoperibolic calorimetry
The equipment is quite adequate for screening purposes. In its simplest form (i.e., a glass tube in an oven), it is a relatively low cost technique that can be assembled with standard laboratory equipment. However, the simple test set-up provides no quantitative thermal data for scale-up purposes, but only T_0 values. The more advanced instruments like the SEDEX and SIKAREX, which are also isoperibolic calorimetry equipment, acquire specific thermal stability data that can be used for scale-up. Furthermore, the small autoclave tests provide gas evolution data.

Because of the operating principles of the equipment, especially in the isoperibolic mode, complex calculation and calibration procedures are required for the determination of quantitative kinetic parameters and the energy release during decomposition. Also, for a reaction with a heterogeneous mixture such as a two-phase system, there may be mass transfer limitations which could lead to an incorrect T_0 determination.

2.3.2 Thermal Stability and Runaway Testing

A principal goal of thermal hazard evaluation is the accurate determination of the thermal stability and runaway behavior of a substance or mixture in combination with gas evolution data. In this section, various thermal stability test methods are discussed. Both isothermal and adiabatic testing techniques are treated.

2.3.2.1 Isothermal Storage Tests

In general, isothermal calorimetry is a very accurate technique in which the heat production of a substance tested under certain conditions is measured as a function of time [109–114]. Performance of these tests at a series of temperatures leads to a quantitative understanding of the relation between the temperature and the heat generation rates of the substance or reaction under investigation. Several apparatus types are available. Most of the equipment is based on the differential measuring principle, which means that both a sample and a reference holder are placed in an isothermally controlled surrounding (e.g., a metal block or liquid bath). Various types of isothermal calorimeters, including their measuring principles, are discussed briefly as follows.

In the isothermal storage test (IST) [115, 116], the heat generated at constant temperature by reacting or decomposing substances is measured as a function of time. An example of an IST is shown in Figure 2.17. The test is

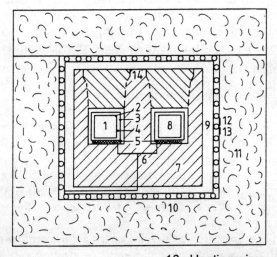

1: Sample
2. Sample vessel
3. Cylindrical holder
4. Air spaces
5. Peltier elements
6. Electrical circuit
7. Aluminum block
8. Inert material
9. Insulation wool

10. Heating wires
11. Glass wool
12. Platinum resistance sensor temperature control
13. Platinum resistance sensor for safety control
14. Platinum resistance thermometer

FIGURE 2.17. Cross-Section of an Isothermal Storage Test (IST).

2.3 PRACTICAL TESTING

applicable to solids, liquids, pastes, and dispersions. The apparatus consists of a large heat sink (e.g, an aluminum block) which is kept at a constant temperature. In the block are two holes, with a heat flow meter at the bottom of each hole. Identical sample holders, one containing the test sample and the other containing an inert substance such as glass powder, are placed on the heat meters. The heat generated by the test sample results in a voltage signal from the heat flow meter that is proportional to the heat flow. Random fluctuations in the heat flow are avoided by monitoring the voltage difference between the two flow meters. Typically, the sample holder, generally made of stainless steel, has a volume of 70 cm^3. The sample mass is about 20 g. Measurements can be performed in the temperature range from −25°C to 150°C. Heat generation can be measured from the lower limit of 5×10^{-3} W/kg to an upper limit of 5 W/kg, with an accuracy of at least ±30% in the lower range to ±5% in the higher range.

Other instruments include the Calvet microcalorimeters [113], some of which can also run in the scanning mode as a DSC. These are available commercially from SETARAM. The calorimeters exist in several configurations. Each consists of sample and reference vessels placed in an isothermally controlled and insulated block. The side walls are in intimate contact with heat-flow sensors. Typical volumes of sample/reference vessels are 0.1 to 100 cm^3, The instruments can be operated from below ambient temperatures up to 300°C (some high temperature instruments can operate up to 1000°C). The sensitivity of these instruments is better than 1 µW, which translates to a detection limit of 1×10^{-3} W/kg with a sample mass of 1 g.

As discussed in Section 2.3.1.2, SEDEX [103, 104] and SIKAREX [106] instruments are also used isothermally. In the case of the SIKAREX, the temperature of the sample is held by a heating coil at constant temperature by establishing a constant rate of heat exchange to the jacket (held about 50 to 100°C below the sample temperature). By measuring the electrical input, a negative copy of the reaction heat profile is obtained. Typical sensitivity of the equipment is 0.5 W/kg operating with a sample size of 10 to 30 g and in a temperature range of 0 to 300°C.

A liquid flow microcalorimeter, the thermal activity monitor (TAM), is commercially available from ThermoMetric (formerly LKB/Bofors). This instrument consists of two glass or steel ampules with a volume of 3 to 4 cm^3 (25 cm^3 ampule available with a single detector), placed in a heat sink block. Recently, an injection-titration sample vessel was developed which acts as a microreactor. This vessel is provided with flow-in, flow-out, and titration lines, with a stirring device. The isothermal temperature around the heat sink is maintained by a controlled water bath. Each vessel holder, containing an ampoule, is in direct contact with a thermopile array, and the two arrays are joined in series so that their output voltages subtract. The two pairs of thermopile arrays are oppositely connected to obtain a differential output, reflecting the difference in heat flow produced in the sample and reference vessels. The temperature range of operation is 10 to 90°C. The sensitivity of this instrument is about 2×10^{-4} W/kg.

Assessment of the results

With the use of isothermal calorimetry, very accurate heat generation rates can be acquired as a function of time. By measurement at several temperatures, global kinetic parameters can be determined, assuming that the reaction mechanism remains the same within the temperature interval investigated. The heat production of the substance under test can be expressed as:

$$\ln q = -\frac{E_a}{RT} + \ln F(Q) \qquad (2\text{-}11)$$

where $[\ln F(Q)]$, in W/kg, is the heat generation factor. From Equation (2-11), it follows that $[\ln q]$ as a function of $1/T$ is a straight line if the relevant q values correspond to an identical degree of conversion of the reaction. Figure 2.18 shows $[\ln q]$ plotted against time as $[\ln t]$ for three isothermal experiments performed at temperatures T_1, T_2, and T_3.

In order to find points of equal degrees of conversion (or equal Q-values) in Figure 2.18, van Geel [115] developed the method to evaluate kinetic data from the so-called isoconversion lines. A heat generating substance that follows Equation (2-11), when stored under isothermal conditions at different temperatures has generated an equal amount of heat (Q) when the product of $t \exp(-E_a/RT)$ has the same value. Thus, for two heat generation/time curves measured at T_1 and T_2, the same amount of heat (Q) has been generated, and thus the conversion is equal when:

$$t_1^{-(E_a/RT_1)} = t_2^{-(E_a/RT_2)} \qquad (2\text{-}12)$$

Denoting the q_1 as the rate of heat generation at T_1 after time t_1, and the q_2 as the rate of heat generation at temperature T_2 after time t_2, then it follows from Equation (2-12) that

$$\frac{q_1}{q_2} = \frac{\exp(-E_a/RT_1)}{\exp(-E_a/RT_2)} \qquad (2\text{-}13)$$

since the times t_1 and t_2 are so chosen that $Q_1 = Q_2$ (equals time of isoconversion). This results in:

$$\frac{q_1}{q_2} = \frac{t_2}{t_1} \qquad (2\text{-}14)$$

Therefore, the heat production as a function of time under isothermal conditions, as recorded at different temperatures, can be expressed as:

$$q \times t = K \text{ (constant) or } \ln q = -\ln t + K \qquad (2\text{-}15)$$

2.3 PRACTICAL TESTING

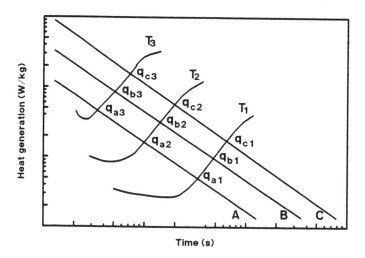

FIGURE 2.18. Rate of Heat Generation (q) of Three Isothermal Experiments as a Function of Time (t) at Three Temperatures (T). [Points of equal conversion of different isothermal experiments at different temperatures are the intersections of the isoconversion lines (A, B, or C) and the heat production lines.]

Equation (2-15) implies that in a ($\ln q$) versus ($\ln t$) plot, points representing equal conversion (or equal amount of heat generation) lie on straight lines which are the isoconversion lines. In Figure 2.18, the points (q_{A1}, t_{A1}); (q_{A2}, t_{A2}); and (q_{A3}, t_{A3}) correspond to the same degree of conversion. The relevant q values of the intersection of the heat generation–time curves with each of the isoconversion lines (the straight lines in Figure 2.18) are plotted in a ($\ln q$) versus ($1/T$) plot in Figure 2.19.

In accordance with Equation (2-11), the E_a is determined from the slope of these lines which is $-E_a/R$.

The isoconversion points (i.e., points of equal Q), can also be determined by numerical integration of the heat production–time curve of each of the individual isothermal tests performed at different temperatures.

By plotting the heat production, q, at points of equal Q (as calculated from each isothermal test) as a function of $1/T$, in the form of Figure 2.19, the E_a can be calculated.

Other kinetic models for determining thermal kinetic parameters are available [90, 117]. Van Geel [116] has also developed a method for determining the safe storage diameter (primarily used for establishing safe storage conditions for propellants) at a given storage temperature.

Using the results acquired from isothermal calorimetry, the safe operating or storage temperature and, if necessary, the required cooling capacity during processing can be established.

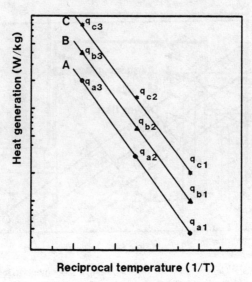

FIGURE 2.19. Rate of Heat Generation as a Function of Temperature at Points of Isoconversion as Derived from Figure 2.18.

Advantages/Disadvantages: Isothermal Calorimetry

A disadvantage of IST measurements is that the experiments take time (days to weeks). Also, several experiments at different temperatures are necessary to get information with respect to the kinetics of the exothermic decomposition. Finally, it may take several hours to reach equilibrium after inserting a sample due to the time-lag of the system. Thus the recorded heat effect may be inaccurate. This is a particular disadvantage in the case of rapid reactions.

Autocatalytic decomposition can be determined by IST techniques. The sensitivity of IST equipment enables measurements to be made at relatively low temperatures, which results in the potential to examine a wide temperature range. This is an advantage because extrapolations outside the temperature range actually examined in tests are reduced. The importance here is that the kinetics of decomposition at high temperatures are not always the same as at the lower temperatures of likely processing.

2.3.2.2 Dewar Flask Testing and Adiabatic Storage Tests

For safety reasons, Dewar flask testing should be carried out in an adequately shielded test facility to protect people and surroundings against a possible explosion as a result of a runaway reaction. In its simplest form, the test unit consists of a Dewar flask that is placed in an oven in order to reduce heat losses to the surroundings, as illustrated in Figure 2.20. The oven is controlled at a desired temperature while the temperature of the contents of the flask is recorded. The Dewar flask must have good insulation properties and the

2.3 PRACTICAL TESTING

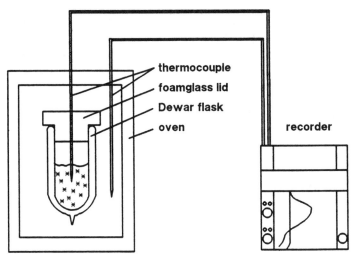

FIGURE 2.20. Simple Test Setup for a Dewar Flask Test.

temperature distribution in the oven should be as uniform as possible. The Dewar flask can be provided with a heater that serves to bring the sample quickly to the desired temperature. The vessel should be closed by a well-insulated lid, since heat losses from the top are important. The material of construction of the lid must be selected to be compatible with any possible corrosive atmospheres. To measure the insulation capacity, a hot liquid is introduced into the Dewar flask standing alone, and the cooling curve is recorded to ambient temperature. The half-life of this cooling curve, which is a measure of the Dewar flask insulation capacity, is thus obtained. Several modifications and applications of Dewar flask testing have been published [28, 118, 119]. Dewar flask testing is applied to: (1) transportation recommendations of the UN [114]; (2) the so-called heat accumulation test; and (3) determining the self-accelerating decomposition temperature (SADT). The latter is used to establish the safe temperature of transportation of reactive substances such as organic peroxides.

Certain equipment configurations allow for the use of Dewar flask testing at elevated pressures. Several arrangements have proved successful such as a sealed glass ampoule in the Dewar flask, a steel pressure vessel in the flask, a Dewar flask in an autoclave under inert gas pressure, and a stainless steel Dewar flask. Dewar flasks provided with an addition line can also be used to study chemical reactions. In Figure 2.21, typical temperature–time curves of Dewar flask experiments are shown.

In some of the equipment, the pressure and temperature are recorded concurrently. The recorded pressure is the result of: (1) the heating of the gas in the head space of the vessel, (2) the vapor pressure, and (3) reaction

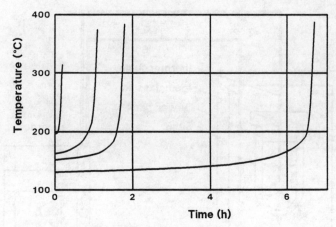

FIGURE 2.21. Typical Temperature–Time Curves of Dewar Vessel Tests (after temperature equilibration between Dewar flask and oven has been reached).

produced gases. With the pressure–temperature vs. time curve, the gas generation of the substance can be calculated (mol gas/mol substance) if sufficient knowledge of the gas solubility in the liquid and the vapor pressure of the sample are available. This calculation may be useful for estimating gas venting requirements for the process.

Modern Dewar flask equipment includes an adiabatic shield, a compensation heater, and a computer to provide for control and for data acquisition and analysis. An example of the application of an advanced design is the adiabatic storage test (AST) [120, 121]. In the AST, the heat generated at nearly adiabatic conditions by the reacting or decomposing substances is determined as a function of time.

Initially, an internal electric heating coil is used to bring the sample to the desired starting temperature. Full adiabatic conditions are approached by keeping the temperature of the oven in which the Dewar flask is situated equal to the temperature of the sample in the test vessel. In this way, the heat loss is kept at a minimum and a nearly adiabatic condition is created by compensation heating. The AST is applicable to solids, liquids, pastes, and dispersions. The capacity of the AST is about 1 liter and the temperature range is about –20°C to 200°C. The arrangement of the AST is shown in Figure 2.22.

Another adiabatic technique [118] uses a liquid as an adiabatic shield. This equipment is controlled by a computer in such a way that the errors that are caused by the heat flow into the heat sink of the sample container are corrected by adding a compensating amount of energy to the system. In this way, no corrections have to be made for calculating the ideal self-heating rate and the induction period using the thermal inertia factor or so-called phi-factor. Dewar flask tests were performed with many unstable organic substances and their adiabatic induction times as a function of start temperatures are given [28].

2.3 PRACTICAL TESTING

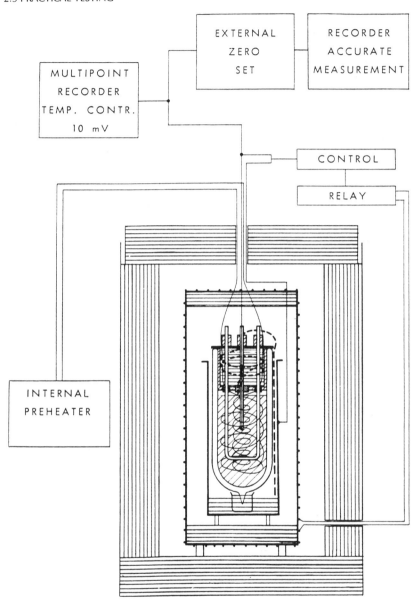

FIGURE 2.22. Arrangement of the Adiabatic Storage Test (AST).

As discussed in Section 2.3.1.2, SEDEX [103, 104] and SIKAREX [106] types of apparatus are also used in adiabatic calorimetric techniques. Compensation for the heat capacity of the sample containment is also a feature. Typical sensitivity of this type of equipment is 0.5 W/kg, the sample size is 10 to 30g, and the temperature range is 0 to 300°C.

Assessment of results

From the temperature–time curve, as recorded in a Dewar flask experiment or in an AST, the heat production as a function of time can be determined. Furthermore, from Dewar flask tests with an accurate internal heater or from AST experiments, the specific heat (C_p) can be determined, or in pressurized and closed vessels, the C_v as well. For the heat production, the following equation holds:

$$q = (m \times C_p + C_{ves}) \times \frac{dT}{dt} \qquad (2\text{-}16)$$

in which C_{ves} is the specific heat of the empty vessel. Equation (2-16) is valid for a completely adiabatic system. In practice, correction factors should be applied for heat loss or heat input. To calculate the C_p (or C_v) value, Equation (2-16) can be transformed to:

$$C_p = \left(\frac{q_m}{m} \times \frac{1}{dT/dt}\right) - \frac{C_{ves}}{m} \qquad (2\text{-}17)$$

in which q_m is the heating power of the internal heater.

An important feature of the adiabatic measuring technique is the determination of the adiabatic induction time, τ_i. The influence of the temperature on the adiabatic induction time is illustrated in Figure 2.23.

For completely adiabatic systems (Biot Number = 0, see Chapter 3), the induction time can be expressed as:

$$\tau_i = \frac{RT_0^2}{E_a} \times \frac{C_p}{F(Q) \times \exp(-E_a/RT_0)} \qquad (2\text{-}18)$$

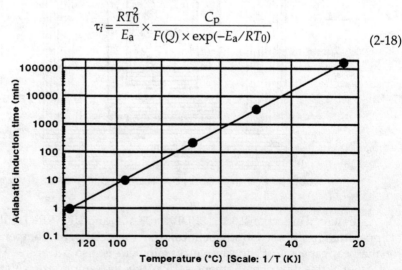

FIGURE 2.23. Adiabatic Induction Time.

2.3 PRACTICAL TESTING

in which T_0 is the start temperature of the substance at which it is initially stored at time zero. Although an estimate of τ_i can be made with Equation (2-18), in practice, the calculation models of Frank-Kamenetskii, Semenov, or Thomas (see Chapter 3) are extremely useful for calculation of τ_i since the system is not completely adiabatic and the Biot Number is not precisely zero.

Advantages/Disadvantages: Dewar Flask and AST
Dewar flask testing and AST are very accurate techniques because a relatively large quantity of substance is used in the test, which makes it possible to get a representative sample. It was found that the results of Dewar flask testing with samples of 200 to 500 cm^3 can be directly applied to reactors with volumes of 0.5 to 2.5 m^3 [122]. Data concerning the induction period of a runaway reaction can be acquired. Furthermore, by testing in a closed pressure Dewar vessel, pressure/temperature data can be obtained. A disadvantage is the time for running these types of tests (up to one month). Also, at a time in a development program when sample availability is limited, a sample size of 0.5 to 1 kg is required. The influence of the phi-factor is less with the sue of the ARC as illustrated previously in Table 2.1 and is discussed in the following section (Section 2.3.2.3).

2.3.2.3 Accelerating Rate Calorimeter (ARC)

The accelerating rate calorimeter (ARC) marketed by Columbia Scientific Instrument Company, uses a measurement technique designed to provide temperature–time and pressure–time data of chemical decompositions or reactions under adiabatic conditions [123–125]. It is particularly well suited for determining a useful onset temperature for exothermic activity. The ARC, illustrated in Figure 2.24, consists of a spherical test vessel with a capacity of 10 cm^3 that contains a solid or liquid for test. This sample holder is mounted inside a calorimeter jacket and is fitted with a thermocouple and a pressure transducer attached through a small tee. In some cases, a thermocouple for directly measuring the sample temperature may be placed inside the sample holder. The jacket is constructed of nickel-plated copper. This jacket contains three themocouples cemented on the inside surface for temperature measurements, and eight heaters to ensure homogeneous temperature distribution. Another thermocouple is attached to the outer wall of the sample holder. The adiabatic conditions are achieved by maintaining the temperature of the sample holder and the jacket exactly equal. As in all other adiabatic measurement techniques, a certain small temperature drift can occur resulting in a slight imbalance of the adiabatic conditions. This drift is minimized by determining and then using an offset voltage (between the jacket and sample holder thermocouples) in the temperature control loop at approximately 50°C intervals over the entire temperature range tested. Measurements are performed by a so-called heat-wait-search operation mode. At first, the sample is heated

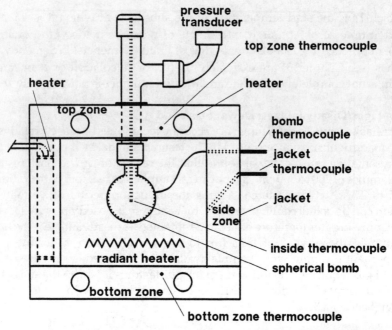

FIGURE 2.24. Accelerating Rate Calorimeter (ARC). (*Note:* Not to scale.)

to the desired starting temperature and held there for a period of time to achieve thermal equilibrium (in general, 10 minutes). Thereafter, a rate search is performed. If the rate, due to the chemical reaction, is less than a preset rate (in general, a calorimetric detection sensitivity of 0.02°C/min is preset), the sample is further heated to the next temperature of heat-wait-search (5 to 10°C higher). This procedure is repeated until a rate greater than the preset rate is detected. Figure 2.25 illustrates the mode of operation. The ARC has a typical sensitivity of 0.5 W/kg. When an exothermic reaction occurs, the temperature difference of the sample holder and the jacket is maintained at zero by a controller.

Assessment of results
According to the literature [77], a process is considered to be low hazard from the thermal standpoint if the normal operating temperature or temperature due to upset is 50°C or more lower than the ARC onset temperature, and the maximum process temperature is held for only a short period of time. However, other factors must be considered in evaluating the thermal hazard of a process such as total enthalpy of reaction or decomposition, potential for reactant accumulation, the boiling point of the reaction mass, and the rate of reaction. The testing must involve all appropriate materials including reactants, intermediates, and products. In some cases, though, the 50°C differential

2.3 PRACTICAL TESTING

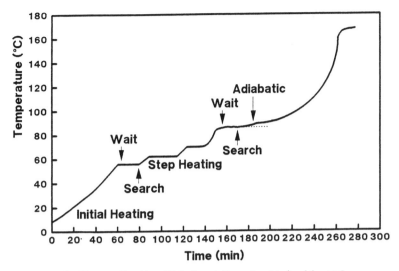

FIGURE 2.25. The Heat–Wait–Search Operation Mode of the ARC.

rule-of-thumb may be conservative for low energy systems or for systems containing large amounts of solvent that can act as a heat sink. If, however, the temperature difference is less than 50°C or the time held at maximum process temperature is considerable (for example, in a dryer), then isothermal aging tests in the ARC or isothermal storage tests (Section 2.3.2.1) are required to detect potentially hazardous exothermic reactions that proceed at such a low rate as to remain undetected in the normal heat–wait–search cycle, or that accelerate with time due to autocatalysis.

Further, the time to maximum rate (TMR) is measured in the ARC, which can indicate the time available for taking defensive or mitigation measures in process upset situations.

From an ARC experiment, the enthalpy of decomposition, ΔH_d, or the enthalpy of reaction, ΔH_r, can be calculated by Equation (2-19):

$$\Delta H_{d \text{ or } r} = \Phi \times m \times C_v \times \Delta T_{ad, s} \tag{2-19}$$

where Φ is the thermal inertia calculated in Equation (2-20) below, C_v is the average specific heat of the sample over the temperature range of test, m is the mass of the sample, and $\Delta T_{ad, s}$ is the experimentally observed adiabatic temperature rise.

$$\Phi = 1 + \frac{(m_{ves} \times C_{ves})}{(m \times C_v)} \tag{2-20}$$

where the subscript "ves" refers to the test vessel or test bomb. Theoretically, if the test vessel absorbs no heat (i.e., the system is completly adiabatic), then $\Phi = 1$. The corrected adiabatic temperature rise of the reaction is ($\Delta T_{ad, s} \times \Phi$).

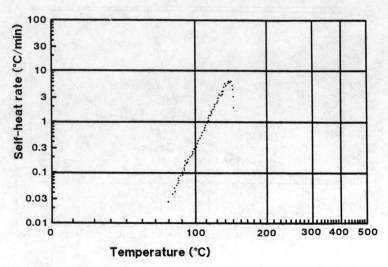

FIGURE 2.26. ARC Plot of Self-Heat Rate as a Function of Temperature.

Figure 2.26 represents an example of an ARC plot of the logarithm of the self-heat rate versus the reciprocal temperature. This graph shows the temperature at which a sample or mixture starts to decompose or react measurably, and the rate at which the sample or mixture liberates heat as a function of temperature. In the ARC experiment represented in Figure 2.26, exothermic decomposition or reaction is first observed at 80°C with a self-heat rate of 0.025°C/min. The maximum temperature reached is 142°C with a maximum self-heat rate of 6°C/min. The data must be corrected for the thermal inertia (Φ) of the system.

Data from ARC experiments can be used to determine the global kinetics [123, 126] of a highly energetic reaction. When the onset is first observed in an ARC experiment, the concentrations have not changed significantly, and therefore:

$$\ln(dT/dt)_{s,o} = \ln[\Delta T_{ad,s} \times c_0^{n-1} \times F(Q)] - E_a/(RT) \qquad (2\text{-}21)$$

The subscript "s" refers to experimental values. The plot of the self-heat rate as a function of the reciprocal temperature (at the start of the reaction) may result in a straight line with a slope of E_a/R which is the zero-order line.

Also, the temperature of no return can be calculated from the data obtained in an ARC experiment [123]. In a given reaction vessel with given heat transfer characteristics, the temperature of no return (T_{nr}) is a metastable temperature such that below T_{nr}, the reaction temperature will not increase since the released heat from the reaction does not exceed the rate of heat removal from the system, and above T_{nr} the reaction temperature will in-

2.3 PRACTICAL TESTING

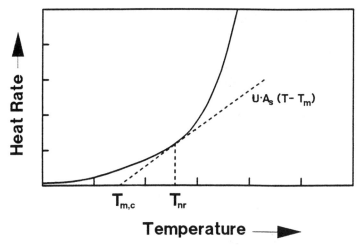

FIGURE 2.27. Heat Release Rate and Heat Transfer Rate versus Temperature

crease, resulting in a runaway reaction. See Chapter 3, Section 3.2.1 for more on the temperature of no return.

Figure 2.27 is an illustration of the capability of heat removal in a system. The temperature of the heat exchanger when reaching criticality ($T_{m,c}$) is determined graphically from the intersection of the straight heat loss line (UA_s) and the abscissa (see Chapter 3, Section 3.1).

Equations for determining the time-to-maximum rate (TMR) and the adiabatic zero-order time to maximum rate at temperature T_{nr} are given in [123].

In addition to the thermal effects of the reaction, pressure data are acquired from an ARC experiment. As in the closed Dewar flask tests, the pressure is the result of : (1) the heating of the free-board gas, (2) the vapor pressure, and (3) the reaction-produced gases. With the pressure-temperature versus time curve, the gas generation of the substance can be calculated (mol gas/mol substance). This is possible if enough knowledge of the gas solubility in the liquid and the vapor pressure of the sample are available. Such a calculation is useful for gas venting estimates for the process.

Advantages/disadvantages: ARC

The ARC is a test instrument that is able to provide information on the runaway behavior of substances and reactions very quickly. Several publications are available regarding the applicability of the results of ARC tests [77, 126, 127–132]. Most of the disadvantages of the ARC discussed are due to the high phi-factor of the equipment relative to plant operating conditions. For example, the phi-factor correction assumes that no additional or different reactions occur at higher temperatures that might be reached under realistic plant conditions. Furthermore, a constant specific heat is generally assumed during the reaction, which directly influences the calculated enthalpy of decomposition or reaction.

Although in determining the onset temperature and the global kinetics in the initial stage of the reaction the , corrections are of lesser importance, these corrections have a dramatic effect on the maximum self-heat rate and the maximum temperature. Care must be taken in the interpretation of such data from an experiment with a phi-factor greater than 1. For direct simulations of plant situations, a phi-factor of 1.0 to 1.05 is used [89].

Another possible disadvantage of the ARC is the relatively small sample quantity used, which thus may not be representative of the full-scale reaction mixture (i.e., scale up from 10 cm^3 to large scale).

ARC experiments are generally run without stirring, although equipment with minimal stirring is commercially available. Since mechanical mixing is usually not appropriate, temperature gradients within the test sample cannot be excluded, nor can the mixing of multiphase systems be readily simulated. Despite the stated disadvantages, the ARC is an extremely useful tool and a recognized industry standard test for the determination of thermal properties.

A comparison of data from ARC experiments and from other testing techniques has been made [126].

2.3.2.4 Stability Tests for Powders

Exposure of powders to air can lead to rapid oxidation, releasing energy in excess of that which would occur from decomposition of the substance alone. The tests described in this section address the hazards associated with the exposure of powders to air during processing. In some cases, the tests are designed to maximize the exposure to air in order to simulate process conditions such as spray drying and fluidized bed drying. Other tests simulate large scale storage of powders.

The stability of powders can be determined by adiabatic storage tests or Dewar flask tests under an air atmosphere (Section 2.3.2.2). Several other dedicated tests have been developed [10, 133–136].

In the Bowes and Cameron test [133], the stability of the powder at constant (uniform) ambient temperature is investigated. Cube-shaped baskets, made of wire gauze, are filled with the substance and placed in an oven that is controlled at the desired constant temperature. The temperature in the center of the cube and of the oven are continuously recorded. By testing at different temperatures, and using a number of cube dimensions, the thermal stability of the powder can be established, that is, the determination of the temperature below which the exothermic decomposition of the powder does not result in a runaway. Bowes [133] has given a number of theoretical calculations for scaling up the test results.

With the United Nations test [134], only two sizes of cubes (one 2.5-cm^3 cube and one 10-cm^3 cube) are tested at 140°C. A sample is recognized as

2.3 PRACTICAL TESTING

susceptible to spontaneous combustion (self-heating eventually resulting in autoignition) if the sample temperature in the center of the cube exceeds 200°C during the 24-hour testing period. However, the criteria to be applied depends upon the practical situation in handling the substance.

Other cubic tests in which a sample is heated in an oven by a controlled temperature program are described in [137].

More rudimentary tests to identify the stability of powders are the so-called "hot plate" tests. Here, the substance is placed in a well-defined manner (circular, specified thickness of the layer, and so forth) on a hot plate with a controlled temperature. The temperature in the center of the layer is recorded continuously, and the progress of the self-heating is followed. Observations are made if self-heating of the substance occurs, the induction period of such self-heating, and the extent of the effect such as smoldering or spontaneous ignition.

Other test methods are the powder bulk, aerated, and layer tests [137, 138]. Several systems in-house built are available. All of these tests operate on the principle that a layer of the substance under investigation is heated in a circulating air oven as the temperature is increased. Air is transported through the sample (in the aerated test, the air flow is downward through the sample), and the temperature of the powder at several places is recorded.

Also, long-term isothermal storage (or low heating rate) tests are used to investigate autocatalytic effects. For example, a sample is held above 100°C for 10 hours to simulate drying operations [137]. In layer tests, the substance layer is heated by hot air passing around it with a fixed velocity.

Assessment of the results

From the results of such tests, the safe handling temperature for powders can be established. It is critical to investigate a representative sample of the material as handled in practice, and consideration should be given to running repeat tests. In practice, fine powdered material will accumulate in the bottom layer of drying equipment or in storage vessels, and may give rise to a thermal runaway since particle size and packing density have significant influences on the oxidative self-heating properties of the substance.

A procedure for definition of safe handling of powders, which is the result of extensive investigations of about 200 samples, is reported in [137] and is summarized as follows:

1. define chemical composition of the material to assure that a representative sample is tested,
2. define the physical characteristics of the material (particle size, packing density, moisture content) to assure that a representative sample is tested,
3. define the powder form in the plant (i.e., bulk material, fluidized material, layer of material) to identify the proper test equipment,

4. define the heat exposure conditions (temperature range, time cycle) in the plant to assure that all reactivity possibilities are tested,
5. examine programmed temperature rise data and, when necessary, do additional isothermal and/or adiabatic testing, which is particularly critical when the substance is heated in bulk form,
6. establish the temperature of decomposition both with air flowing through and supplied by diffusion, and
7. define maximum plant temperature conditions using appropriate safety factors.

According to the report, since the introduction of this procedure, some 300 powder materials have been dried in approximately 5000 drying operations with no major runaway decompositions.

2.3.3 Explosibility Testing

Specific explosibility tests are discussed in this section. It is emphasized that detonation tests in particular and a number of deflagration tests are of the type that must be run by experts in specialized facilities.

2.3.3.1 Detonation Testing
There are, in general, two types of detonation tests [10, 22, 24]. In one type of test, the propagation of a detonation shock wave is determined. Examples of methods for this test are the BAM 50/60 test [139], the TNO 50/70 steel tube test [120], and the USA GAP test for solids and liquids [140, 141].

In the other type of test, the strength of the detonation (explosive power) is determined. Examples of methods for this type of test are the lead block test [139] and the ballistic mortar test [141]. Only the first type of test, which determines the possibility of a detonation, is discussed here.

Propagation of a detonation wave depends on a number of physical parameters:

- the quantity of the sample—a detonation propagates better within a larger quantity; the diameter of the substance under investigation should exceed a minimum value, which depends on the substance (in some cases, the diameter of the applied tube tests is too small);
- the degree of confinement (rigidity of the wall)—at a high degree of confinement, the detonation propagates better;
- the specific density—the velocity of the detonation wave increases with increased specific density of the substance, although some explosives (e.g., ammonium nitrate) show a certain maximum specific density above which no propagation occurs;

2.3 PRACTICAL TESTING

- the strength and shape of the initiation source—the more powerful the initiation shock wave, the more likely a detonation wave will propagate;
- the temperature;
- the presence of entrained bubbles or of cavitation; and
- the particle size and crystal structure.

In general, detonation test apparatus consists of a steel tube that is filled with the substance under investigation. One end of the tube is provided with a booster charge consisting of an electric detonator covered by detonative material. The other end is either closed or provided with a witness plate. One type of steel tube apparatus is provided with a velocity probe to record the shock wave velocity as shown in Figure 2.28.

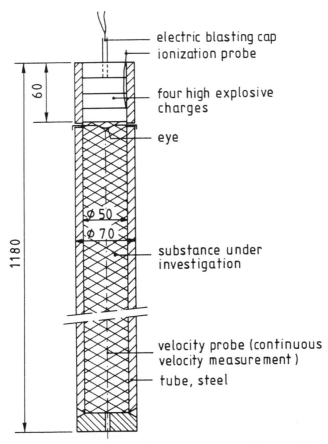

FIGURE 2.28. Test Set-up of the TNO 50/70 Steel Tube Test (dimensions in mm).

Tests must be performed in a bunker. The booster charge is detonated and, after the event, observations are made of the fragmentation pattern of the tube and/or the condition of the witness plate.

Assessment of results
In the steel tube test, the fragmentation pattern is a measure of whether or not a substance detonated, for example, a substance is capable of detonation if the tube is completely fragmented. The velocity of the reaction front is also a measure of detonation, that is, a detonation has occurred if this velocity is higher than the velocity of sound in the substance. When a witness plate is used, a substance is said to be capable of detonation if either a hole is punched through the plate or the tube is fragmented along its entire length.

2.3.3.2 Deflagration Testing and Autoclave Testing
A deflagration is a chemical reaction that propagates in the material at subsonic velocity by thermal energy transfer [29.142]. A deflagration can exhibit either very mild propagation of a reaction front (1 to 10 mm/min) or very rapid propagation (10 to 1000 mm/min) [29, 143]. For example, some fertilizers show very mild deflagrations, whereas some organic peroxides, di-nitro and tri-nitro compounds (at higher temperatures), and oxidizer/reducer mixtures show very rapid deflagrations resulting in violent action such as very significant pressure rises. The severity of the deflagration is not only related to the type of substance, but is also dependent on the magnitude of confinement, for example, the initial pressure [144], and the temperature of the bulk of the substance mass.

Deflagration tests run under ambient pressure are relatively rudimentary. They provide information concerning only the propagation rate of deflagration after forced initiation. Examples of these tests are the UN deflagration test [143], dedicated to classification of organic peroxides, and the UN trough test [145], dedicated to classification of fertilizers.

The UN deflagration test consists of a Dewar vessel with a volume of about 400 cm^3. The vessel is filled with preheated material (standard temperature is 50°C if the stability of the substance permits), and the substance is initiated at the top of the vessel with a flame. The propagation of deflagration is recorded by temperature sensors that are located in the substance at preset distances. From the time required for passing two temperature sensors and from the known distance between them, the deflagration velocity can be calculated.

In the trough test, the sample (only solids) is introduced in a horizontal wire mesh cage with an inner volume of 11 liters. The substance is initiated at one end of this trough by a gas burner or electrical heating source and the propagation of the deflagration front is established and noted.

Different types of equipment are available for the investigation of deflagration properties of substances under physical confinement or under elevated pressures.

2.3 PRACTICAL TESTING

One type of apparatus is designed to study the pressure effect of a runaway reaction that results in a deflagration (i.e., to study the thermal explosive behavior of the chemical system). In this type of test, the strength of the deflagration (explosive power) is determined. Sometimes, substances or reaction mixtures that will not deflagrate but show only runaway reactions are studied in the same equipment in order to obtain pressure–time and temperature–time data. Suitable test equipment is based on a closed autoclave system, with or without a vent. Either the substance, or the entire autoclave system if relatively small, is heated until runaway occurs, followed by, if possible, spontaneous ignition of deflagration [31, 120, 143, 146–149]. Measurements of the internal pressure of the autoclave and, in some equipment, the temperature of the substance are recorded. Working pressures of most of the autoclaves vary from 5 to 1500 bar. With more advanced autoclave systems, data on deflagration velocities and/or decomposition kinetics during the runaway stage are obtained.

Other types of autoclaves are designed to investigate deflagration after initiation by an igniter. Examples of this equipment are the time/pressure test [143, 150], designed primarily to classify organic peroxides, the Strand Burner type of apparatus (e.g., Crawford-type bomb used in investigations of deflagrative properties [24, 147, 151–153]), and the constant pressure autoclaves [24, 31, 154].

An example of the influence of pressure on the deflagration rate is shown in Figure 2.29 as obtained in testing with the constant pressure autoclave (CPA). An organic peroxide, t-butylperoxybenzoate (TBPB), was tested at several temperatures and pressures. It is clear from the data that the deflagra-

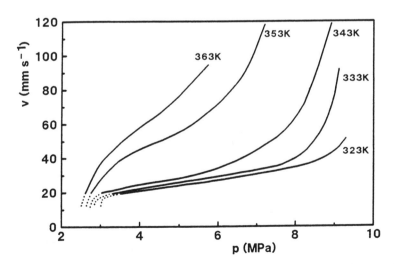

FIGURE 2.29. Deflagration Rate of TBPB at Different Temperatures as a Function of Pressure Established in the CPA.

tion rate increases with both increases in the initial temperature and the initial pressure. These phenomena create a significant hazard in storing or handling substances subject to deflagration under either confined conditions are at elevated temperatures.

Some substances show a deflagration-to-detonation transition (DDT). This hazardous phenomenon can occur when a number of conditions are encountered, including: (1) the substance itself is able to detonate (Section 2.3.3.1), the substance is sufficiently confined, and (3) sufficient quantity of the substance is available. This DDT can be the result of confinement as the pressure increases during the deflagration. The mass transfer from the propagation zone (escaping gases) accelerates rapidly in a definite direction, which results in compression waves. The compression waves gradually increase into shock waves which are characteristic of a detonation. These shock waves overtake the deflagration waves. A DDT is especially likely to occur in long pipes or in pipelines containing substances prone to deflagrate/detonate.

The UN DDT tests [22] are used to investigate this phenomenon. These tests are, in principle, similar to the tube tests described in Section 2.3.3.1, but as an initiation source, a hot nichrome wire, or a nichrome wire surrounded by black powder, placed in the center of the tube is used instead of a detonation booster.

Assessment of results

The results obtained from deflagration testing are either pressure–time data or deflagration rates as a function of pressure and/or temperature.

With pressure–time data from spontaneous deflagrations (thermal explosions), the maximum expected pressure, and the time for pressure-rise can be estimated. Furthermore, the so-called "specific energy" (F) or "explosive power" of substances [24, 31] by Equation (2-22) from experiments in which the sample mass is varied.

$$F = \frac{p_{max} V}{m_0} + K$$

(2-22)

where F is the specific energy (force constant), p_{max} is the maximum pressure recorded, V is the volume of the autoclave, m_0 is the initial sample mass, and K is a constant, all in consistent units. Equation (2-22) is valid when most of the substance is in the gaseous state at the moment of reaching p_{max}, and the pressure is less than 250 bar. By plotting V/m_0 versus $1/p_{max}$, F can be determined from the slope of the line. Using the calculated value of F, an approximation of the maximum pressure rise after thermal explosion of the substance in other containment volumes (V) can be made.

Deflagrations can be initiated by external stimuli such as shock, friction, or sparks.

A spontaneous deflagration is a consequence of internal heating (runaway). The runaway, on crossing a certain temperature limit, results in spon-

2.3 PRACTICAL TESTING

taneous initiation of a reaction wave that passes through the mass, hence a deflagration. Both natural convection in the reaction mass and heat transfer properties (e.g., hot spots) affect the time and place of initiation. Spontaneous deflagrations, which normally appear at higher temperatures, are usually unacceptable in any normal processing. Therefore, protective measures should be taken, such as adding inerts or diluents to the substance, or tight control of the temperature to exclude a runaway leading to deflagration (see thermal stability testing, Section 2.3.2). In most cases, it is impossible or certainly impractical to vent the gases evolved during a spontaneous deflagration because a large amount of gaseous products are evolved in a short period of time.

Control of a deflagration after initiation by a source such as a hot spot, a flame, or a spark, depends on the rate of deflagration, the confinement, and the accumulation of heat from the evolved energy. Very slow deflagrations can sometimes be controlled under nonconfined situations. Under confined conditions, pressure builds up with simultaneous energy accumulation, which increases the deflagration velocity, most likely to an unacceptable level in processing.

2.3.3.3 Mechanical Sensitivity Testing

It is important to know the sensitivity of substances to mechanical stimuli with the subsequent capability of propagating into a deflagration or a detonation (explosive substances). In practice, a substance can be wedged between surfaces, for example, between a container and its lip or between flanges, which will impact on the substance. The maximum temperature increase due to friction between nonmelting surfaces can be as high as 2000°C. This may lead to initiation of a deflagration or detonation for sensitive substances. In fact, for substances which are sensitive to impact, dropping may be sufficient to lead to such initiation.

Mechanical sensitivity testing is divided into sensitivity to mechanical shock, also called sensitivity to impact, and sensitivity to friction. Equipment is available to investigate these properties [10, 22, 24, 140, 155, 156].

Determination of friction sensitivity is applicable to solids, pastes, and gel-type substances. To determine the friction sensitivity, a thin sample is placed under a load between two roughened surfaces, and the surfaces are then rubbed together in a controlled manner. The load can be varied. Results from this action, such as smoke, cracking, or discoloration, are observed. Examples of apparatus of this type are the BAM friction apparatus, shown in Figure 2.30, the rotary friction test, and the ABL friction test.

The determination of the mechanical shock sensitivity or impact sensitivity is applicable to both solids and liquids. The principle involved is that a drop weight falls from a specified height onto the confined test sample. The load can be varied by changing the height of the drop and by changing the drop weight.

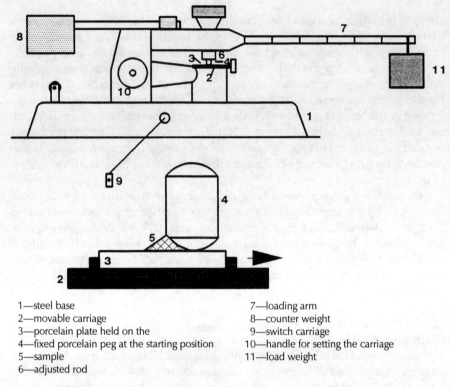

1—steel base
2—movable carriage
3—porcelain plate held on the
4—fixed porcelain peg at the starting position
5—sample
6—adjusted rod
7—loading arm
8—counter weight
9—switch carriage
10—handle for setting the carriage
11—load weight

FIGURE 2.30. The BAM Friction Apparatus: Horizontal and Vertical Cross-Sections

Observations are made concerning sample decomposition or explosion (e.g., a noisy report, smoke, or fire), cup deformation, and possible gas generation.

Examples of equipment for mechanical shock or impact testing are the BAM Fallhammer, the Rotter test, the thirty kilogram Fallhammer test, and the Bureau of Mines Impact Apparatus [45]. The latter is shown in Figure 2.31.

Assessment of results

Two types of results are obtained from mechanical sensitivity tests: (1) no reaction, or (2) decomposition with or without an explosion. The magnitude of friction and the impact sensitivity reported is the smallest load at which a positive result has been noted. The objective of mechanical sensitivity testing is to establish whether or not the substance is sensitive under normal handling conditions. However, this objective may not be reached. The test results may not truly reflect process conditions because most testing is carried out at ambient temperature and pressure. Since results are dependent on the type of test apparatus used, the interpretation of the results for use in practical applications requires much experience in this field of testing.

2.3 PRACTICAL TESTING

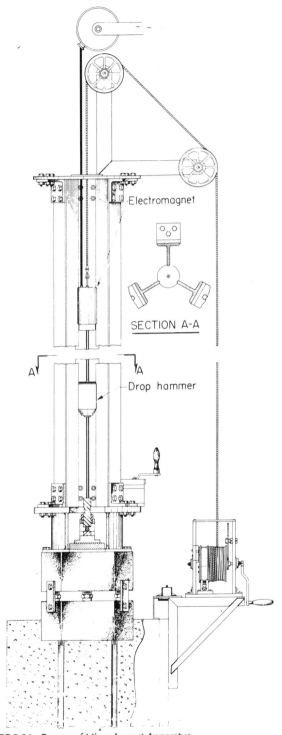

FIGURE 2.31. Bureau of Mines Impact Apparatus.

2.3.3.4 Sensitivity to Heating under Confinement

When large quantities of a substance are handled, sensitivity of the material to heating under confinement may need to be considered to demonstrate the effect on the stored/handled, and probably confined, substance in the event of an external heat load. Tests such as the steel sleeve test or Koenen test [24, 137], the Dutch pressure vessel test (DPVT) [143], and the United States pressure vessel test (US-PVT) [143] may be applicable. These tests are used mostly for transportation considerations. The tests generally subject the sample substances to very high energy inputs under confined conditions, and thus are more severe than the deflagration and autoclave tests previously discussed in Section 2.3.3.2. As an example, the Koenen test, used mainly in Europe, is illustrated in Figure 2.32.

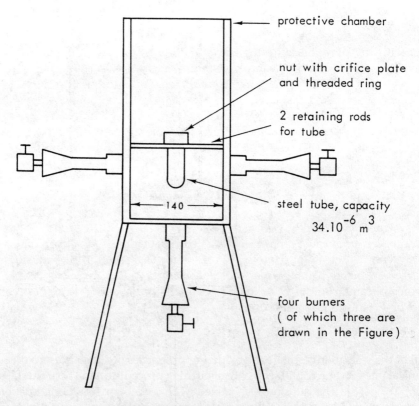

FIGURE 2.32. Set-up of the Koenen Test.

2.3.4 Reactivity Testing

2.3.4.1 Pyrophoric Properties

Very rudimentary test have been described to study whether or not a material is pyrophoric [134, 135]. In general, the tests involve a procedure in which 1 to 2 cm^3 of powder or 5 cm^3 of liquid is poured from a height of about 1 m onto a noncombustible surface. Observations are made as to whether or not the substance ignites during dropping (powders) or within 5 minutes (all substances). The test is repeated six times.

In another test with liquids, 0.5 cm^3 is delivered from a syringe onto dry filter paper, and observations are made to see if ignition or charring occurs on the paper within 5 minutes. This test is repeated three times unless a positive reaction is observed earlier.

Assessment of results
When a substance ignites in one of the tests, or, with liquids, chars the filter paper in a test, the substance is considered to be pyrophoric.

2.3.4.2 Reactivity with Water

As with the pyrophoric testing, a rudimentary test has been described to investigate the potential for gas generation or reaction of solids and liquids with water [134, 135]. A 25-g sample (either solid or liquid) is introduced into a closed conical flask that has both a dropping funnel containing water and a suitable gas volume metering device. The stopcock of the dropping funnel is opened to bring water into contact with the sample substance. If gas is evolved, the amount is recorded as a function of time, and, if the identity of the gas is unknown, it is tested for flammability.

The heat generation on contact with water must also be considered. Here, one of the calorimetric tests described previously in this chapter may be used.

Assessment of results
When a material causes rapid energy release on contact with water, particularly with the release of flammable gases that could result in an explosion, it is recognized as a dangerous substance (as discussed in Section 2.2.4.3). Protective and cautious handling procedures should be used to exclude contact of the substance with water.

2.3.4.3 Oxidizing Properties

All of the tests to investigate the oxidizing properties of substances involve a conical pile or a horizontal strip type of burning procedure and apparatus [134, 136, 157, 158]. The substance is mixed with a known dried combustible material such as sawdust, cellulose, or sugar, in various ratios. The burning velocity of a horizontal strip or the burning time of a conical pile of the mixture is then measured as a definition of the oxidizing properties of the substance. The pile can be ignited by a gas flame or a hot platinum wire.

Also, the spontaneous ignition temperature for liquid or volatile oxidizers can be investigated by testing [157]. Here, a predetermined quantity of sawdust (12 to 50 mesh) is added to a reaction vessel and brought to the desired test temperature. The liquid oxidizer is then cautiously injected with a long hypodermic syringe into the vessel. The extent of reaction is determined from continuous temperature measurements and by visual observations.

Assessment of results

In all of the test methods, the observed burning rates or times are compared to the burning rates or times of reference substances under the same experimental conditions. Based on the comparison with several oxidizing substances having a known classification, the hazard of the sample substance is rated. It should be emphasized that some oxidizer/combustible mixtures will react vigorously. Some mixtures are able to detonate, depending on the specific composition (see Section 2.2.4.2).

2.3.5 Flammability Testing

The well-known flashpoint is a measure of the flammability of liquids.

The flammability of solid substances is determined by burning rate tests [10, 134, 135]. From a mold, a pile of the substance under investigation is placed on a noncombustible, nonporous, and low heat-conducting base plate. One end of the pile is ignited by a hot gas flame or a hot platinum wire (temperatures above 1000°C). The burning rate is established and measured.

Assessment of results

A substance is recognized as having flammable properties when the burning rate exceeds a certain value. For example, according to the UN classification, a solid substance is recognized as having flammable properties if the burning rate exceeds 2.2 mm/s using the UN test method [134].

3
CHEMICAL REACTIVITY CONSIDERATIONS IN PROCESS/REACTOR DESIGN AND OPERATION

3.1 INTRODUCTION

As shown in Chapter 1, the design of a safe chemical plant is based on three principal parameters: the available energy in the chemicals involved, the rates of reaction/decomposition, and the plant processing system. Decomposition with the attendant heat release, having roots in both the "chemicals" and "reaction rates" characteristics, is a major factor in hazard evaluation. The potential energy and the acquisition of relevant data by theoretical and testing procedures regarding this energy were discussed in Chapter 2. The potential heat release rate and the aspects relating to scale (application and impact on design and operation) are now reviewed in Chapter 3.

Here, the technical aspects of the desired reactions are discussed, including the determination of the reaction enthalpies, the heat capacity of the reaction mass, kinetic data, and the rates of temperature and pressure rise. The test methods by which this information is obtained are described in Section 3.3.2, and the results from these tests are reviewed in Section 3.3.3. The appropriate information is usually obtained on a small scale, varying from a few cubic centimeters to a few liters. In general, the results of these tests have to be reviewed from the perspective of scale; thus, scale-up is of great importance and is treated in Section 3.3.4. The integration of all of the information for process design is covered in Section 3.3.5. Then follows Section 3.3.6, which is concerned with the storage and handling of reactive materials from the viewpoint of potential hazards.

An important theme continuing throughout Chapter 3 is the concept of inherent safety. In general, the scope of all safety efforts is to reduce the probability and severity of hazardous incidents and upsets. Measures aimed at decreasing the probability of an incident are called preventive, while those aimed at reducing the severity of an incident are called protective or mitigating. Achieving inherent safety by proper design is, of course, preventive. As Regenass [15] has pointed out, "It is obvious that inherent safety must be sought at the basic conception of a process." It follows also that the process should be continually reevaluated for hazards, especially before scale-up and before making any changes during commercial plant operations. During design of the commercial plant as well as during commercial operations, the evaluation of any process changes and then proper management of such changes are critical factors.

The first step in approaching inherent safe design is to identify and understand basic thermodynamic data. A systematic approach to inherent safety, as outlined throughout this book, requires information concerning:

- the energy content of the system,
- the exotherm of both desired and undesired reactions,
- the reaction rates,
- the potential consequences of a runaway reaction, and
- the definition of credible upset scenarios.

3.1.1 Thermal Hazards: Identification and Analysis

This section discusses how a runaway reaction occurs and lists some of the process deviations that can lead to such a runaway. Equipment for identifying potentially hazardous process steps is reviewed, and general principles for inherently safe process design are given.

Additional hazards in plant operations include specific gas and dust explosions, but such considerations are outside the scope of this book. Reference can be made to another CCPS Guidelines book [159] for issues relative to certain explosions.

3.1.1.1 Cause, Definition, and Prevention of a Runaway

Runaway reactions can be triggered by a number of causes, but, in most cases, their resultant features after initiation are similar [31]. Whenever the heat production rate exceeds the heat removal rate in a reaction system, the temperature begins to rise and can get out of control. The runaway starts slowly but the rate of reaction accelerates, and the rate of heat release is very high at the end. Most runaways occur because of self-heating with the reaction rate (and reaction heat output) increasing exponentially with temperature, while the heat dissipation is increasing only as a linear function of the temperature.

The effect of a runaway can be neglected only if the amount of heat available in the reaction mass at the point of runaway is low even at high

3.1 INTRODUCTION

temperature. Problems can arise if the heat from the desired reaction can raise the temperature to a level that other exothermic reactions start, such as decomposition or polymerization, thus causing pressure increase, rapid boiling of solvent, or production of gases. The production of gases from decomposition or boiling can lead to an increase in pressure and loss of containment. If enough energy is available in the system, the runaway may initiate a deflagration.

Regenass [160] mentions the factors, as listed below, that can lead to a runaway. When any two or more of these factors are present, there is a potential for thermal runaway:

- high heat release of intended reaction,
- high heat release of potential decomposition,
- high heat release of competing reactions,
- accumulation of reactants or intermediates,
- insufficient heat removal,
- thermally hazardous materials involved,
- too high a temperature, and
- loss of solvent (heat sink).

There is a potential thermal runaway upon the combined occurrence of two or more of the above listed factors. For example, an accumulation of reactants in combination with insufficient heat removal leads to a runaway of the desired reaction. The resulting temperature increase (now uncontrolled) may lead to an explosive decomposition if other exothermic reactions, such as decompositions or polymerizations, occur within the range of the temperature increase.

A self-heating reaction will become an uncontrollable runaway whenever the heat generation rate exceeds the heat removal rate. In Figure 3.1, these rates are plotted against reactor temperature for the case of a well-stirred, cooled reactor system. The heat generation rate, which is proportional to the reaction rate, is an exponential function of the absolute temperature (Arrhenius equation) and appears in Figure 3.1 as a curved line. The heat removal rate, which is proportional to the difference of reactor temperature and coolant temperature, is thus a linear function of temperature and appears as a straight line. The intersection of this straight line with the x-axis is the temperature of the cooling medium (T_m).

Heat balances occur at the intersection of the heat generation curve and the heat removal line (points C and D). Stable operation will occur at point C. A reaction temperature lower than point C will result in self-heating up to point C because the heat generation rate exceeds the heat removal rate. At temperature T_B, the heat removal rate exceeds the heat generation rate, so the reaction temperature will fall until point C is reached. Although point D is a heat balance point, no stable operation is possible here; a temperature slightly lower than that at point D will result in a decrease in reactor temperature to

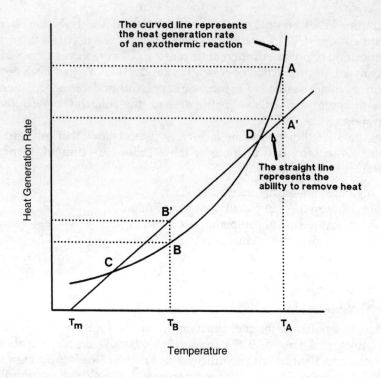

FIGURE 3.1. Typical Heat Generation and Heat Removal Rates as a Function of Temperature.

that at point C, whereas a slightly higher temperature will result in a runaway. The importance of point D is that the reactor temperature here is the critical temperature above which a runaway is inevitable. At temperature T_A, for example, the heat generation rate exceeds the rate of heat removal and the reaction will develop into a runaway situation.

Other aspects of the heat balance are discussed more extensively in Section 3.2 and are shown on Figure 3.7 in that section.

Now that some of the possible causes of a runaway reaction have been discussed, the question of preventive measures remains. The type of information needed can be grouped according to the triangle discussed in Chapter 1 (Figure 1.1). These parameters are as follows:

1. Available Energy in the Chemicals Involved—
The thermochemical evaluation of the reactants is the place to start, as discussed in Chapter 2. If the potential energy release is 250 cal/g of substance (1000 J/g) or more, it is necessary to determine initiation properties, such as a sensitivity to impact and friction. Thermochemical evaluation of reaction

3.1 INTRODUCTION

mixtures and intermediate products should be carried out as well because of possible hazards in processing and storage. The crude and pure products must be evaluated in a similar fashion.

2. Enthalpies and Rates of Reaction/Decomposition—
The enthalpy of reaction that is most needed is the not the enthalpy of any specific reaction, desired or undesired, but rather the global or macro enthalpy of reaction at various conditions, including different temperatures. This term is defined as the heat evolved by the reaction system in which reactants are converted into products and by-products by one or more reactions. The global enthalpy of reaction is difficult to calculate, but relatively easy to measure by experiment. Any such experiment must simulate the conditions which exist in the process under study (i.e., concentrations, temperatures, and pressures). The experimental values will, of course, include the heat evolved from the desired reaction(s) and from all of the undesired reactions that happen to occur under the selected conditions.

Besides the overall enthalpy of reaction, the rate of heat evolution at various temperatures is needed in order to design a process. It is desirable, though, whenever possible, to have a complete understanding of the kinetics of all of the reactions and to know the heat contributed by each. Extending the temperature of experimental measurements far above the desired and anticipated reaction temperature will give information about additional reactions that can occur if the reaction at lower temperature is allowed to run away. It is also desirable to get data on reaction streams with various concentrations of reactants.

Experimental data can be obtained from the DSC and from reaction calorimeters for the conditions of the desired reactions, and from the DSC, the ARC, the Reactive System Screening Test (RSST—Fauske and Associates) and from the Vent Size Package (VSP) for conditions allowing undesired reactions. The pressure effect can be studied using the ARC or DIERS methods. From the results of these tests, the rate of temperature rise and the maximum acceptable conditions for specific equipment can be calculated. The same holds for the pressure rise rate.

From the heat capacity of the reaction mass, preferably determined in relatively large units such as in bench-scale trials, and the total heat effect, the adiabatic temperature rise for the global reaction in question can be estimated and/or predicted.

The global rates of heat generation and gas evolution must be known quite accurately for inherently safe design.. These rates depend on reaction kinetics, which are functions of variables such as temperature, reactant concentrations, reaction order, addition rates, catalyst concentrations, and mass transfer. The kinetics are often determined at different scales, e.g., during product development in laboratory tests in combination with chemical analysis or during pilot plant trials. These tests provide relevant information regarding requirements for cooling capacity and reaction control. Such information is indispensable for scale-up to plant facilities.

Investigation of the global rates of reaction can be carried out in instrumented bench-scale equipment. such as the RC1 (Mettler-Toledo) plus on-line chemical analysis. Commercially available equipment allows well-controlled process conditions, and can be used in a variety of modes (e.g., isothermal, adiabatic, temperature programmed). The test volumes, which may be up to 2 liters depending on the energy involved, enable reasonable simulation of process conditions, and are more representative than very small samples, particularly for mixed phase systems. The scale of such equipment permits the collection of accurate data.

3. *Plant Process System Facilities*—
A distinction can be made between scale-up of equipment and scale-up of test results. An aspect of equipment scale-up is shown by the following example. The maximum heat generation during a reaction determines the required cooling capacity of the plant unit. In scaling, the volume increases by the cube of the vessel diameter, but the heat transfer area increases only by the square. Therefore, the ratio of heat production to heat removal becomes considerably larger at a larger scale which would result in correspondingly higher temperatures and possibly a runaway if design corrections are not made. External cooling systems or internal cooling coils may have to be included in the plant scale equipment. Scale-up of laboratory test data must be reviewed carefully as well.

Critical heat production rates (i.e., heat production rates that still do not lead to a runaway), are often determined by small scale experiments. However, the effect of scale-up on these rates, as discussed in [161], must be taken into account. An indication of the effect of scaling in an unstirred system is shown in Figure 3.2. In this figure, the heat production rate (logarithmic scale) is shown as a function of the reciprocal temperature. Point A in the figure represents critical conditions (equivalent heat generation and heat removal) obtained in a 200 cm^3 Dewar vessel set-up. It can be calculated from the Frank-Kamenetskii theory on heat accumulation [157, 162] that the critical conditions are lowered by a factor of about 12 for a 200 liter insulated drum. These conditions are represented by

line B. Taking into account a conservative activation energy of 40 to 50 kJ/mol (a value typical of self-heating materials such as coal and lignites), this means that the critical temperature in the case of the 200-liter drum is estimated to be about 90°C lower than the critical temperature of the Dewar vessel (60°C instead of 150°C). In cases with uncomplicated kinetics, good data from a representative sample of the process material in a calibrated DSC instrument can be used to indicate the hazardous temperature regions for a larger scale. In the curves shown in Figure 3.3 [163], the enthalpy of reaction found by integration of the DSC scans and the first-order kinetic constants found by

3.1 INTRODUCTION

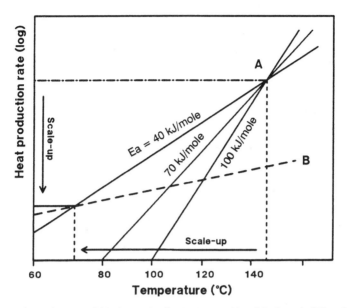

FIGURE 3.2. Relation between Critical Heat Production Rates of Small Scale and of Plant Scale (small scale = 200 cm^3 Dewar vessel, point A) (large scale = 200 liter drum, line B)

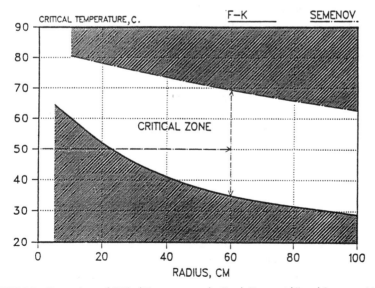

FIGURE 3.3. Comparison of Critical Temperatures for Frank-Kamenetskii and Semenov Models (Right Cylinder Configuration).

plotting the peak temperatures against the scan rates as in ASTM E-698 [164], were used to generate curves of runaway temperature versus the radius of the container for both the Semenov model (good agitation) [165] and the Frank-Kamenetskii model (no agitation values used in the plot were:

Enthalpy of reaction: 300 cal/g
Activation energy: 30, 000 cal/gmol
Log_{10} (Frequency Factor): 15 (factor unit, min^{-1})
Sample thermal conductivity: 0.0003 cal/ (sec) (°C) (cm^2)
Container heat transfer coefficient: 0.0005 cal/ (sec) (°C) (cm^2)

Both curves show the decrease in runaway temperature as the radius is increased. The plot also indicates that the reaction run at 50°C will not run away in a vessel of 10 cm radius, or probably 20 cm radius, as these points are below both curves.

When the reactor is scaled up to 60 cm radius, however, the operating point is between the two curves. This means that the reaction can be safely run at 50°C in a well-agitated process vessel of 60 cm radius with the heat transfer coefficient as stated above becauseerating point is below the Semenov curve. In case the agitation is lost, however, the Frank-Kamenetskii curve becomes the better predictor of runaway temperatures, and because the operating point is above this curve, the estimate is that the reaction will run away. The calculation of the Frank-Kamentskii method is available in ASTME-1231 [166].

It is clear that proper and appropriate test methods must be used to obtain the necessary data for scale-up.

3.1.1.2 Some Simple Rules for Inherent Safety

The risk of an incident is based on potential severity (or consequence) and probability of occurrence. Both aspects should be thoroughly considered in design and operation for inherent safety. *Potential severity* can be reduced by:

1. *keeping potential excursions within acceptable limits, for example,*
   - by avoiding accumulation of reactive components, and
   - by diluting the reactive components, thus introducing a higher latent heat and/or heat capacity.
2. *keeping inventories low*
   - by minimizing reactor size, and
   - by avoiding storage of potentially hazardous materials in the synthesis train.

The probability can be reduced, for example,

- by running well within limits of safe operation, and
- by providing sufficient cooling capacity in case of emergencies (however, in the case of an autocatalytic decomposition, it may be too late to control the emergency solely by cooling by the time the runaway is recognized).

3.1 INTRODUCTION

Severity and probability of upsets may thus be eliminated or mitigated if information is available concerning:

- the total energy contents of the components and of the system,
- the respective reaction energies of both desired and undesired reactions,
- the reaction rate as a function of temperature, and
- the effect on temperature in case of cooling failure.

3.1.1.3 Strategy for Inherent Safety in Design and Operation

A scheme to evaluate a plant design for inherent safety is shown in Figure 3.4. The basic data required for such an evaluation are shown in the top box. The enthalpy of reaction and the specific heat of the reaction mass determine the maximum increase in temperature that may occur if all the heat that is released is accumulated in the reaction mass (i.e., at adiabatic conditions). The activation energy, the reaction rate constant, and the enthalpy of reaction are essential parameters in defining the rate at which heat can be generated as a function of the operating temperature. These parameters, in fact, determine the heat removal capacity that is required in order to avoid undesired temperature increases of the reaction mass.

An increase in temperature is not dangerous in itself as long as it can be kept under control. Temperature rises result in a nondesired runaway only if secondary reactions are initiated or if temperature control becomes impossible. Cooling can be obtained in several ways such as by the use of specifically designed cooling systems or by reflux systems. For a cooling system, heat transfer characteristics and mass flow of the coolant can be adapted for the required cooling capacity. For a reflux system, a solvent is introduced into the reaction system. The boiling point of the solvent should be equal to or lower than the maximum allowable temperature in the reactor. Essentially all of the heat generated by the on-going reaction can then be removed by the latent enthalpy of vaporization of the boiling solvent. The solvent is subsequently condensed in a heat exchanger on top of the reactor and returned as reflux to the reactor. Basic characteristics and requirements for safe operation are, in the reflux case, the properties of the solvent or diluent (boiling temperature, enthalpy of vaporization), the quantity of solvent (which depends on the maximum evaporation rate), the condensing capacity of the reflux system (sufficient evaporated diluent must be condensed), and the rate of return flow of the condensed solvent. Verification that sufficient diluent is present before starting the reaction is important even though additional solvent/diluent may be necessary in the process as the reaction proceeds.

Pressure can increase as a result of a runaway. Gas can be generated as a normal reaction product, as a result of evaporation of low-boiling components in the reaction mass, or as a by-product produced during a runaway. Gas can be produced at very high rates at continued runaway conditions. Maximum

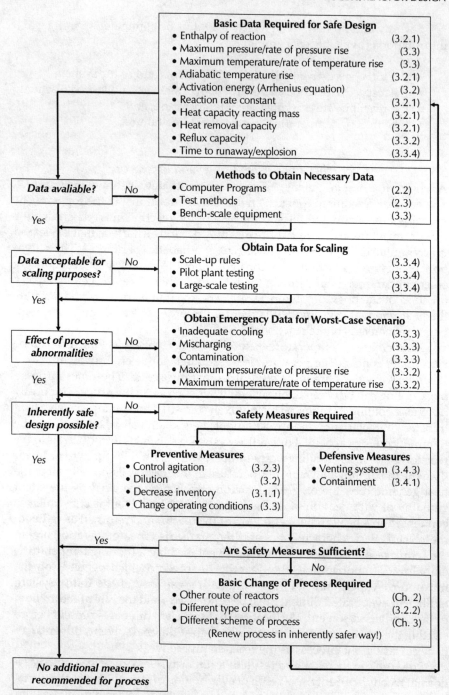

FIGURE 3.4. Process Hazard Evaluation Scheme.

3.1 INTRODUCTION

gas production and production rates under operating and emergency conditions must be known for inherently safe design of the reaction vessel.

Time-to-runaway, described later, is an important parameter in the development of adequate emergency measures. Short times require the use of well-defined operating modes or require completely automated controls.

The data must be reliable. This implies that data only from the literature or by calculations either must be regarded as completely acceptable by experts in the field, or must be checked experimentally. Moreover, the data must be acceptable for scale-up to plant equipment. Emergency requirements for the worst credible scenarios must be determined. For example, HAZOP studies are appropriate here. If all the data and analyses show that no runaway is to be expected during operation, the design may be regarded as inherently safe, which implies that no additional measures need to be introduced from this perspective. If the data and analyses show that a runaway is possible but the reactor vessel can be designed and built to contain the worst credible case, the design may be regarded as inherently safer than one requiring venting facilities.

In most cases, however, the design will not be inherently safe so that measures and/or controls will have to be introduced into the design. Two types of safety measures are defined: (1) preventive, and (2) defensive (protective or mitigating).

A measure is preventive if it prevents the occurrence of a runaway, a decomposition, or a hazardous secondary reaction. The system conditions remain close to operating conditions. Excessive increases in temperatures and/or pressures are avoided. Preventive measures include feed rate control systems, interlocks to prevent the reaction from starting unless sufficient diluent is present or the cooling system is working, and tests for the presence of catalysts or unwanted impurities. Preventive measures are always to be preferred over protective or mitigating (defensive) measures.

Defensive measures are directed at reducing the consequences of a runaway. In most cases, the increases in pressure are the major problem during an undesired event. Damage to the reactor follows if the pressure exceeds the design pressure of the vessel. As this is obviously to be avoided, most defensive measures are directed at keeping potential pressure increases within defined acceptable limits. A typical defensive measure is the application of vent systems with a capacity sufficient to keep the pressure below a preset limit. Venting implies the transport of material outside the reaction system. Generally, the material should not be released directly to the environment, especially if the chemicals involved are of a hazardous or toxic nature. Thus, the addition of a vent system implies the introduction of supplementary equipment in the vent line hat permits absorption or treatment of the vented substances. No further steps are required if the introduced safety measures sufficiently prevent the occurrence of unsafe situations. However, if unsafe situations will occur, and the risks cannot be managed, then a re-design of the process is essential.

3.1.1.4 Equipment to be Used for the Analysis of Hazards

The need for experimental thermodynamic and kinetic data is clear by now. The equipment designed to provide this information for the chemicals involved are described in Chapter 2, and include the DSC, DTA, ARC, Sikarex, SETARAM C-80, and DIERS technology. Kinetic data for the desired reaction are preferably obtained with instrumented bench-scale equipment such as the RC1. This type of equipment is discussed in Section 3.3.

Thermodynamic data (enthalpy of reaction, specific heat, thermal conductivity) for simple systems can frequently be found in date bases. Such data can also be determined by physical property estimation procedures and experimental methods. The latter is the only choice for complex multicomponent systems.

3.2 REACTOR, HEAT AND MASS BALANCE CONSIDERATIONS

Safe operating conditions are obtained by successfully managing the energy content in the system at all times. Safe conditions can be defined by analysis of the system starting from the heat balance (temperature related) and the mass balance (pressure related). Not only the inherent energy of the system, but also the kinetics (i.e., the energy release rate), are of great importance in determining temperature and pressure increases. The effects of heat generation and reaction mass composition are discussed in this section.

3.2.1 Heat and Mass Balances, Kinetics, and Reaction Stability

The fundamental law of the conservation of energy leads to the following heat balance for well-defined systems:

$$C_P \times \underset{(I)}{(dT/dt)} = \underset{(II)}{Q_{in}} - \underset{(III)}{Q_{out}} + \underset{(IV)}{Q_{reaction}} \qquad (3\text{-}1)$$

The terms of Equation (3-1) are heat flows, that is, flows of energy per unit time (per unit of mass). The accumulation of energy in the system and the corresponding temperature rise per unit time can be calculated if the terms (II), (III), and (IV) are known. These terms comprise the following types of energy:

Term (I):
—heat accumulation of system.
Term (II)
—thermal heat input by either external or internal heating,
—heat input by reactants added to the system, and
—mechanical heat input, e.g., by agitators, pumps, etc.
Term (III)
—heat output by outflow of products or vapors,

3.2 REACTOR, HEAT AND MASS BALANCE CONSIDERATIONS

—heat exchange with external cooling ($q = UA_s \, \Delta T$), and
—heat loss by radiation.

Term (IV) (heats/enthalpies):
  —enthalpy of reaction of desired, side, and consecutive reactions,
  —enthalpy of solution (e.g., the enthalpy of solution of concentrated sulfuric acid in aqueous systems is substantial),
  —enthalpy of fusion, and
  —enthalpy of crystallization/precipitation.

3.2.1.1 Adiabatic Temperature Rise

The thermal hazard of a chemical system is always determined by its potential to show an increase in temperature and in the production of gases. If no heat is exchanged, i.e., under adiabatic conditions, the resulting rise in temperature is known as the adiabatic temperature rise. Such an extreme condition may occur at loss of cooling, loss of agitation, or with rapid decomposition reactions. Because of their relatively large mass, plant-scale vessels operate close to adiabatic conditions unless heat transfer systems are installed and running. The adiabatic temperature rise is a good measure of the hazard since it is the maximum possible increase. The adiabatic temperature rise of a system (ΔT_{ad}) can be derived directly from the heat balance as shown in Equation (3-1) since the terms (II) and (III) are zero in an adiabatic system. The adiabatic temperature rise (ΔT_{ad}) at complete conversion of the reaction mixture, assuming that the contents are homogeneous and physically independent of temperature, is:

$$\Delta T_{ad} = \frac{c_R \times \Delta H_r}{\rho \times C_p} \tag{3-2}$$

where c_R is the reaction concentration in mols/unit volume. The relation between the temperature and the conversion in an adiabatic reactor is shown in Equation (3-3), according to the heat balance:

$$T = T_0 + (\Delta T_{ad} \times X_A) \tag{3-3}$$

where X_A is the degree of conversion.

It is obvious that reducing c_R (i.e., increasing the dilution), results in a reduction in the adiabatic temperature rise and, thus, can help to keep the reaction temperature within acceptable constraints. The global heat balance over the system, with all heat generation terms included, is required to obtain the actual adiabatic temperature rise. From the safety perspective, the adiabatic temperature rise is a useful design parameter, although it must be emphasized that it shows only a maximum effect and not a rate.

Reactions in a system with a high ΔT_{ad} may lead to a high reactor temperature and may, for example, boil off all of a solvent diluent. As a consequence, organic materials may decompose into small, gaseous molecules, which will result in an increase in pressure. Consider, for example, an organic material with a C_p of 2 kJ/(kg)(°C) and an enthalpy of decomposition, ΔH_r, of 1800 kJ/kg. According to Equation (3-2), the ΔT_{ad} is about 900°C (without correcting for density). This usually means an extensive decomposition will occur. In such a case, it is important to know the rate at which heat is released and the time at which the extreme condition is reached. Therefore, Townsend [132] used the term "time-to-explosion" or "time-to-runaway," an expression introduced by Semenov. The time-to-runaway, which can be easily calculated for adiabatic conditions, gives a conservative (short) estimate of the time available for applying corrective measures in the progression to a runaway.

3.2.1.2 The Reaction

From the reaction rate r, the reaction enthalpy ΔH_r, and the reaction volume V (= reaction mass/density), the heat production per unit of time (q) can be calculated:

$$q = r \times V \times \Delta H_r \tag{3-4}$$

The effect of a runaway follows from Equation (3-4). It shows that a system is more hazardous if it has a high reaction rate, a large inventory, or a high enthalpy of reaction and/or decomposition. If one of these three parameters is reduced and controlled, q may be kept under control as indicated in Figure 3-5.

3.2.1.3 Reaction Rate

The reaction rate for a reaction involving materials A, B, . . . , D, is often approximated by an expression of the following type [168.169]:

$$r = k_n(c_A^a c_B^b \cdots c_D^d), \quad (a + b + \cdots + d) = n \tag{3-5}$$

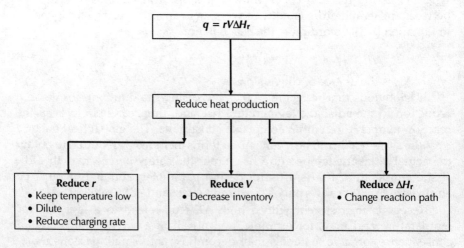

FIGURE 3.5. Methods to Reduce the Heat Production q.

3.2 REACTOR, HEAT AND MASS BALANCE CONSIDERATIONS

in which $a, b, \ldots, d$ may be, but are not necessarily, related to the stoichiometric coefficients. The exponents in Equation (3-5) are the order of the reaction. Thus, the order is ath order with respect to A, bth order with respect to B, and nth order overall. Reaction orders may vary with temperature and other variables. The reaction rate constant k is characterized by the overall order n of the reaction and, therefore, is often identified as k_n. The effects of the reaction rate constant and the concentration of reactants on the reaction rate are treated in the following sections.

3.2.1.4 Reaction Rate Constant

The reaction rate constant, k, is an exponential function of the reciprocal of the absolute temperature and is defined by Equation (3-6), the Arrhenius equation [169, 170]:

$$k = A \exp\{-E_a/(RT)\} \tag{3-6}$$

A typical graph of k as a function of temperature is shown in Figure 3.6. The increasing slope shows the importance of determining a maximum allowable temperature in process equipment so that the heat removal capacity is not exceeded. Under adiabatic conditions, the temperature will reach the calculated maximum only if the reactants are depleted. The actual maximum temperature in a system with some heat dissipation will, of course, be somewhat lower than the calculated value.

In a typical example as shown in Figure 3.6, the rection rate constant increases rapidly with increasing temperature as described in Equation (3-6). It follows, then, that it is necessary to determine the maximum allowable temperature in the system. This is the maximum temperature at which heat

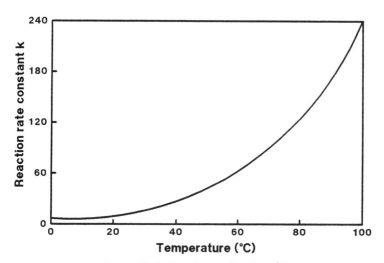

FIGURE 3.6. Reaction Rate Constant k of a Reaction as a Function of Temperature.

from the main reaction and any side reactions (e.g., decompositions), can be removed safely by available cooling.

3.2.1.5 Concentration of Reactants

In addition to the reaction rate constant, the concentration of reactants influences the reaction rate significantly as shown in Equation (3-5). For example, assume that the reaction A → P is a first-order reaction with respect to A. Then, from Equation (3-5), the reaction rate becomes $r = kc_A$. For a concentration twice as high, the reaction rate increases by a factor of two. Dilution, then, is a method to lower the reaction rate and to moderate the increases in temperature and pressure. Dilution results in a lower final pressure provided the vapor pressure of the diluent is relatively low.

In heterogeneous systems, another phenomenon is encountered. Here, the concentration of reactants can be established by the mass transfer. Mass transfer is increased if the contact area between the two phases (the interface) is increased, which, for example, can be done by increasing the rate of agitation. Any reduced agitation results in a smaller contact area between the two phases, which could decrease the reaction rate. In other situations, vigorous agitation may be preferred, for example, to avoid accumulation of reactants through increasing the reaction rate. Increase in mixture solubility of reactants because of a product of the reaction must be considered as well.

3.2.1.6 Effect of Surrounding Temperature on Stability

The effect of temperature of the reactor surroundings on the reaction stability can be discussed via the simplified heat-flow model [171–173] shown in Figure 3.7, which is really an extension of Figure 3.1. Consumption of material is not considered in this figure.

The heat balance for a batch reactor, where Q_{in} is zero, shows that heat accumulation is the difference between heat production and heat removal, from Equation (3-1), leading to Equation (3-7):

$$C_p(dT/dt) = Q_{reaction}(T) - Q_{removal}(\Delta T) \qquad (3\text{-}7)$$

Equation (3-7) is valid under ideal conditions. This means that the temperature of the reaction mass is assumed to be uniform throughout the reactor. The heat generation ($Q_{reaction}$) is an exponential function of temperature and is represented by the curve in Figure 3.7.

The heat removal depends linearly on the difference between the reactor temperature and the coolant temperature since $q_m = UA_s(T - T_m)$, where the subscript "m" refers to the cooling medium. The heat removal is represented by straight lines on the figure. The heat flow is zero if no heat is removed, which is the case if the coolant temperature is equal to the temperature of the system. Thus, the intersection of a heat removal line with the x-axis (e.g., $T_{m,1}$)

3.2 REACTOR, HEAT AND MASS BALANCE CONSIDERATIONS

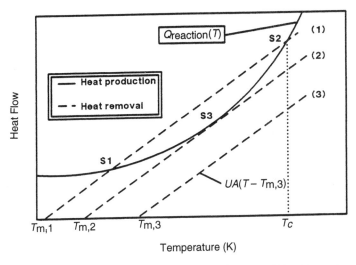

FIGURE 3.7. Stability as a Function of Heat Production and Heat Removal.

represents the coolant or ambient temperature which, if reached by the system, results in zero heat exchange.

A stable situation is represented by the heat removal line (1), provided the temperature of the system is lower than the temperature T_c, which here is equal to the temperature of no return (T_{nr}). Under these conditions, the cooling capacity of the system exceeds the heat that is generated and the system temperature will invariably drop to the point of intersection S1.

Theoretically, the system is still stable exactly at temperature T_c (intersection point S2). In practice, however, the system becomes unstable with only a very small deviation of T_c. At a fractionally lower temperature, the system will go to stable point S1. But at a fractionally higher temperature, the system will generate more heat than can be removed. Consequently, the temperature of the system will increase, the reaction will runaway, and a hazardous situation will likely occur depending on the mass that is still reactive. At any point of intersection of the heat production line with a heat removal line, the heat production equals the heat removal as indicated below in Equation (3-8).

In heat removal line (2), the heat production line and the heat removal line have only one point of intersection (S3). Here, a critical situation exists. In practice, S3 is not a stable point for operation because a small temperature increase will lead to a runaway reaction. The point S3 is of interest, however, for the calculation of the maximum ΔT that can be used for safe cooling of a batch reactor. At S3, the reaction temperature T and the ΔT are at maximum values and the slopes of the two curves are equal. Equation (3-9) and Equation (3-10) are true and valid at point S3. Substituting the value of $UA_s(T - T_m)_{max}$

in Equation (3-9) for the equivalent exponential term in Equation (3-10) gives Equation (3-11), which reduces to Equation (3-12). For a reaction with E_a of 18 kcal/mol at 100°C, the maximum ΔT is about 15°C. The hazards of running batch reactions with a coolant at too low a temperature are not widely known or considered, but certainly should be.

In heat removal line (3), no intersection occurs with the heat production line. This situation always leads to runaway. Thus, the temperature of the surroundings (e.g., coolant) may not exceed $T_{m,2}$ in Figure 3.7 in order to keep the system stable.

The appropriate equations relative to the discussion of Figure 3.7, for a zero-order reaction, follow:

$$q = rV\Delta H_r = UA_s(T - T_m) \tag{3-8}$$

$$A \exp\{-E_a/(RT)\} \times c_0 V\Delta H_r = UA_s(T - T_m)_{max} \tag{3-9}$$

$$UA_s = A \exp\{-E_a/(RT) \times c_0 V\Delta H_r \times \{E_a/(RT^2)\} \tag{3-10}$$

$$UA_s = UA_s(T - T_m)_{max} \times \{E_a/(RT^2)\} \tag{3-11}$$

$$(T - T_m)_{max} = (RT^2)/E_a \tag{3-12}$$

where T_m is the ambient or heating/cooling medium temperature.

3.2.1.7 Effect of Agitation and Surface Fouling on Stability

The stability of a reactive system is effected by agitation and by fouling of the reactor side of the heat exchange surface as illustrated in Figure 3.8 for the case in which mass transfer is not limiting.

The heat removal from a reaction system is a linear function of the overall heat transfer coefficient U, which in itself is a function of the rate of agitation as well as the coolant flow rate and the thermal conductivity of intermediary metal or insulating layers. Consequently, an increase in agitation speed may result in an increase in the slope of the heat removal line. In Figure 3.8, the effect that this has on the stability can be defined. At a coolant temperature $T_{m,1}$ marginal conditions occur for operating by heat removal line 2. An increase in the rate of agitation can result in a new heat removal line 1, thus increasing the stability of the system. A decrease in agitation rate may result in heat removal line 3, where insufficient heat is removed and a runaway will result.

Some examples in which this agitation effect is more likely to occur are reactions during which the viscosity changes significantly, such as in polymerizations, and reactions with suspensions. Equipment dimensions, type of agitators, and type of solvents and coolants used affect the heat transfer as well [174].

3.2 REACTOR, HEAT AND MASS BALANCE CONSIDERATIONS

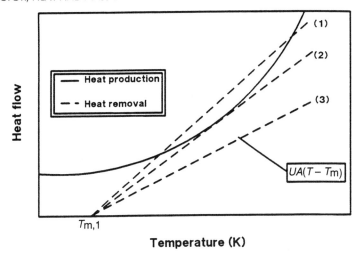

FIGURE 3.8. Effect of Agitation and Surface Fouling on Heat Transfer and Stability.

Since the overall heat transfer coefficient U depends also on the coolant flow rate, it must be emphasized that loss of coolant flow or fouling of the heat transfer surface on the coolant side has a similar effect as shown for loss of agitation.

Fouling of heat transfer surfaces affects the slope of the heat removal line. An increasing degree of reactor side surface fouling can change the heat removal line from the slope shown on 1 to that shown on 2, and then ultimately to that shown on 3, which is an unstable situation. Fouling of the internal reactor surface often occurs if solid particles are formed during the reaction.

3.2.1.8 Mass Balance

The mass balance for a system of reaction component X is formulated by [171, 172, 175, 176]:

$$\left\{\begin{array}{c}\text{accumulation}\\\text{of mass X}\end{array}\right\} = \left\{\begin{array}{c}\text{inflow of}\\\text{mass X}\end{array}\right\} - \left\{\begin{array}{c}\text{outflow of}\\\text{mass X}\end{array}\right\} + \left\{\begin{array}{c}\text{reacted}\\\text{mass X}\end{array}\right\} \quad (3\text{-}13)$$

The mass balance can be used to calculate the pressure rise in a reactor from a reaction by the use of the ideal gas law or the Van der Waals equation [47]. The evaporation of solvents is another cause for a rise in pressure, and, in fact, the evaporation of volatiles is a major factor to be controlled. As an example, the vapor pressure of acetone at three temperatures, as calculated from Weast [47], is given in Table 3.1. The example shows that an increase in temperature of 100°C leads to an increase of pressure by a factor of 23.

A third reason for a pressure increase is gas production from decomposition of one or more reactants, which may also be calculated by a review of mass balances.

TABLE 3.1
Vapor Pressure of Acetone at Different Temperatures

Temperature, °K	Vapor Pressure, bar
300	0.32
350	1.97
400	7.19

The maximum pressure increase due to desired and undesired reactions is, in practice, often investigated in laboratory equipment such as the ARC, RSST, and VSP (see Chapter 2, and Section 3.3 below) to verify the results from the mass balance. These test show the pressure history for a set of extreme conditions and, therefore, provide an estimate for gas production which can be used in scale-up calculations. Limitations in the applications of such tests may occur because of the test size relative to the plant scale. These aspects are discussed in Section 3.3.4. Such tests must be carried out for the complete range of temperatures which may occur on the plant scale.

3.2.2 Choice of Reactor

During the development of a chemical process, a choice must be made regarding the type of reactor to be used on a plant scale. Some theoretical considerations and their practical impact on reactor issues are presented here. Choosing the right type of reactor can indeed improve the safety of the process. The considerations are reflected as well in the mode of operation. Reactors are characterized by type of operation (i.e., batch, semi-batch, and continuous).

A batch reactor is an agitated vessel in which the reactants are precharged and which is then emptied after the reaction is completed. More frequently for exothermic reactions, only part of the reactants are charged initially, and the remaining reactants and catalysts are fed on a controlled basis; this is called a semi-batch operation. For highly exothermic reactions and for two-phase (gas–liquid) reactions, loop reactors with resultant smaller volumes can be used.

Batch (noncontinuous) reactors are characterized by nonstationary conditions, that is, there are composition and heat generation changes during operation.

In continuous reactor systems, all reactants are continuously fed to the reactor, and the products are continuously withdrawn. Typical continuous reactors are stirred tanks (either single or in cascades) and plug flow tubes. Continuous reactors are characterized by stationary conditions in that both heat generation and composition profiles remain constant during operation (provided that operating conditions remain unchanged!).

3.2 REACTOR, HEAT AND MASS BALANCE CONSIDERATIONS

Reactors can be operated in different modes, such as:

- adiabatic,
- isothermal or pseudo-isothermal by heat exchange,
- isothermal by evaporation/reflux,
- nonisothermal or temperature programmed, or
- combinations of above modes.

The combination of modes is often used in practice. Frequently, the temperature is controlled by reflux operation with sufficient condenser cooling capacity as indicated in Section 3.1.1.3. Here, a solvent is used with a boiling point at or slightly below the desired operating temperature. The heat required to evaporate the solvent compensates the heat generated in the reaction, thus maintaining a constant operating temperature. A runaway may be prevented in this mode of operation since an increase in heat generation is compensated for by an increase in the solvent evaporation rate, provided that sufficient reflux condenser capacity and solvent are available. A significant increase in the reaction temperature is thus avoided.

The choice of a reactor is usually based on several factors such as the desired production rate, the chemical and physical characteristics of the chemical process, and the risk of hazards for each type of reactor. In general, small production requirements suggest batch or semi-batch reactors, while large production rates are better accommodated in continuous reactors, either plug flow or continuous stirred tank reactors (CSTR). The chemical and physical features that determine the optimum reactor are treated in books on reaction engineering and thus are not considered here.

The potential thermal hazard for a process may be decreased by choosing the right type of reactor. The hazard implications in each of the four reactor types listed below are given in Table 3.2:

- Plug Flow Reactor (PFR),
- Continuous Stirred Tank Reactor (CSTR), also called the back-mixed reactor,
- Batch Reactor (BR), and
- Semi-Batch Reactor (SBR).

The applicability of these reactor types is discussed on the basis of a few typical examples. The choice of a reactor type should be made with the objectives of avoiding hazardous undesirable reactions, maximizing the selectivity (yield) of the desired product, and achieving a high production rate. Frequently, the choice of the reactor type to avoid hazardous undesired reactions that may occur along with the desired reaction is the same choice that would be made to maximize the yield and production rate of the product. In this section, brief comments are made on some of the principles for obtaining selectivity and for reducing hazards. The subject is considered in depth in books on reaction engineering [167, 168].

TABLE 3.2
Comparison of Different Reactor Types from the Safety Perspective

PFR	CSTR	BATCH	SEMI-BATCH
ADVANTAGES			
• Low inventory • Stationary condition	• Stationary condition • Agitation provides safety tool • Streams may be diluted to slow reaction	• Agitation provides safety tool	• Controllable addition rate • Agitation provides safety tool • Large exotherm controllable
DISADVANTAGES			
• Process dependency • Potential for hot spots • Agitation present only if in-line mixtures are available • Difficult to design	• Large inventory • Difficult to cool large mass • Difficult start-up and shutdown aspects • Precipitation problems • Low throughput rate	• Large exotherm difficult to control • Large inventory • All materials present	• Starting temperature is critical (if too low, reactants will accumulate) • Precipitation problems

1. *Single Reactions*—For all reactions of orders above zero, the CSTR gives a lower production rate than the batch, semi-batch, or kinetically equivalent plug-flow reactor.

2. *Multiple Reactions*—Choosing a reactor type to obtain the best selectivity can often be made by inspection of generalized cases in reaction engineering books. A quantitative treatment of selectivity as a function of kinetics and reactor type (batch and CSTR) for various multiple reaction systems (consecutive and parallel) is presented in [168].

3. *Two Reactions, Consecutive*—In the case of a series of two first-order reactions, $A \rightarrow R \rightarrow S$, the selectivity (yield) of intermediate R will generally be higher in the batch, semi-batch, or plug-flow reactor than in the CSTR. The ratio of the rate constants for the first and second reactions is important. When k_1/k_2 equals, say, 10, the reactor type has a very large effect on the yield of intermediate, but when k_1/k_2 equals, say, 0.1, the effect of reactor type is considerably less and almost disappears at high conversion. When the final product S is the desired product and the intermediate R is a hazardous material, using a CSTR will give the advantage of keeping the concentration of R low.

4. *Two Reactions, Parallel*—In the case of two parallel first-order reactions, $A \rightarrow B$ and $A \rightarrow C$, where only one of the products, B, is desired, the CSTR is

3.2 REACTOR, HEAT AND MASS BALANCE CONSIDERATIONS

preferred over the batch, semi-batch, or plug-flow reactor. Also, the semi-batch mode is better than the batch mode.

5. *Two Reactions, Different Orders*—In the case of a desired second-order reaction and an undesired first-order reaction, such as $A + B \rightarrow C$ and $A \rightarrow D$, where C is the desired product, the batch, semi-batch, or plug-flow reactor is preferred.

6. *Two Reactions, Effect of Temperature*—When the activation energy of the desired reaction is greater than the activation energy of the undesired reaction, increasing the reaction temperature will provide better selectivity for the desired product. When the activation energy of the desired reaction is less than that of the undesired reaction, decreasing the reaction temperature will provide the better selectivity. A table with the best choices for reaction temperature for several complex reaction schemes is given by Levenspiel [168].

An example of the effect of temperature on selectivity (yield) for the case of two reactions where A goes to product P by a first-order reaction, and P goes to impurity X by a second-order reaction is shown in Figure 3.9. Say that the undesired reaction is highly exothermic. If the product P is removed as soon as it is formed, the second (undesired) reaction will not occur. It is evident that the overall reaction would be more hazardous and the yield of product P less if an incorrect reactor type is selected. From Figure 3.9, it can be seen that the higher the temperature, the greater the decrease in selectivity. At low

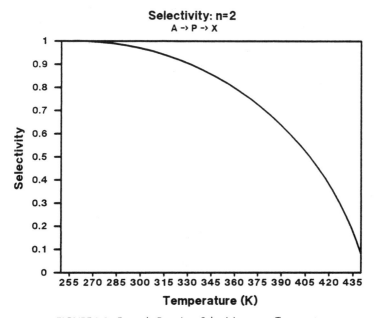

FIGURE 3.9. Example Reaction: Selectivity versus Temperature.

selectivity, more heat will be generated per mol of reaction mixture (due to the highly exothermic consecutive reaction) than at high selectivity. The temperature will rise faster and a self-accelerating process will occur.

Another way to improve selectivity with respect to product P, then, is to select the right reaction temperature. If the activation energy for the desired reaction ($E_{a,1}$) is higher than the activation energy for the formation of X ($E_{a,2}$), a higher reaction temperature is recommended. If $E_{a,1} < E_{a,2}$, a falling reaction temperature is recommended which can be achieved with variable cooling rates.

As another example, consider the reaction A + B yielding desired product P, which can then decompose at a high temperature in a highly exothermic manner to form the undesired substance X. The theoretical adiabatic temperature rise of both reactions can be reduced by dilution, as mentioned previously in Section 3.2. However, this will reduce the reactor productivity so that the amount of diluent should be chosen to be as small as practical for economic reasons. If all the reactants A and B react within a short period of time, the temperature will rise rapidly and thus the formation of X can take place relatively rapidly, which increases the temperature rise per unit of time. It is likely that the decomposition temperature may be reached with dangerous consequences. Thus, it is important to minimize the amount of X formed by controlling the temperature and reducing the accumulation of reactants. This may be done by choosing a semi-batch reactor (e.g., feeding B to A). Accumulation, and hence the potential temperature rise for the first reaction, will be reduced by reducing the feed rate.

7. *Batch Reactor versus Semi-Batch Reactor*—It is frequently necessary to run a rapid reaction that is highly exothermic. Even without detailed calculations, intuition suggests it is prudent to withhold one reactant from the initial charge and add it to the reactor gradually at a controlled rate to keep the heat evolution under control. This semi-batch reaction mode is often used for the stated purpose. Two calculations to be made are the adiabatic temperature rise and the maximum safe ΔT of the reaction temperature above the cooling medium. The first calculation is to determine whether or not a runaway batch reaction will be hazardous (likely), while the second is made to determine if a practical cooling rate can be designed per Equation (3-12) and the preceding discussion.

It should be noted that there are cases in which some selectivity will be lost in choosing a semi-batch mode over a simple batch reactor. If the desired product decomposes by a consecutive reaction, the yield will be higher in the batch reactor [177]. If, on the other hand, the reactants are producing by-products by a parallel reaction, the semi-batch process will give the higher yield. In any case, if the heat production rate per unit mass is very high, the reaction can then be run safely under control only in a semi-batch reactor.

As an important note, it can be dangerous in a semi-batch reactor to choose the starting temperature of the reaction too low (see Figure 3.10, line $T_0 = T_1$),

3.2 REACTOR, HEAT AND MASS BALANCE CONSIDERATIONS

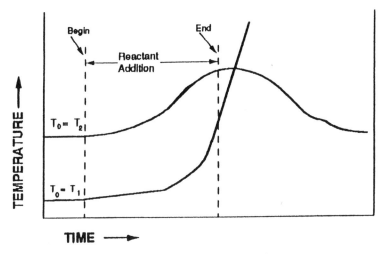

FIGURE 3.10. Effect of T_0 in a Semi-Batch Reactor.

because the reaction must proceed at a reasonable rate to prevent the accumulation of reactants. When reactants reach a critical concentration, the reaction may start generating heat at a higher rate than that which can be removed and controlled, thereby allowing the temperature to rise above the decomposition temperature. If a higher T_0 is chosen (line $T_0 = T_2$), the reaction is sufficiently rapid to be controlled at all times. There is, of course, an upper limit for T_0; it must be chosen at a point sufficiently below the temperature where decomposition will initiate and proceed rapidly.

Increasing the temperature may affect the selectivity adversely, but if it is acceptable for yield and selectivity reasons, it is preferred to a decrease in feed rate, which decreases the reactor productivity.

The rules for intrinsically safe batch and semi-batch reactor operations are extensively discussed by Steensma [175] and Steinbach [177, 178].

3.2.3 Heat Transfer

As described in Section 3.2, a key method to control the temperature of a reaction mass is to make sure that the transfer of heat to the cooling system is sufficient and that the system is properly designed. Frequently, this means the presence of good agitation in the reaction vessel not only to provide mixing of the reactants to control and sustain the reaction, but also to provide for good heat transfer control. Not all reaction vessels can be readily agitated, however, for example, plug-flow reactors. Some background information is presented here regarding heat transfer in both agitated and nonagitated vessels.

3.2.3.1 Heat Transfer in Nonagitated Vessels

It is not practical to stir all reaction systems, for example, bulk polymerizations, postpolymerization reactions, fixed-bed catalytic reactors, and plug-flow reactors. Although multipoint temperature sensing is often used as a key solution to determine a runaway in nonagitated vessels, the occurrence of hot spots may not always be detected.

In a related case, if the agitation is interrupted in an exothermic reacting system, the removal of heat from the mass is difficult; thus, the danger of a runaway occurs. Due to free convection, criticality will occur in the upper layers of the fluid. A typical practical example is in nitration reactions, where loss of agitation causes the acid inorganic layer to separate from the organic reaction mass leading to a serious runaway situation. with the likely evolution of significant by-product gases. Another practical case of an nonagitated system involves substances reacting under storage conditions.

A criterion to avoid thermal decomposition in nonagitated cases in the top layer of the core (the likely hot spot) of a vertical cylindrical vessel is to assure that the maximum temperature difference between the reacting bulk fluid and the bulk fluid of the coolant (perhaps the ambient air) is always less than shown in Equation (3-14) [173, 179]:

$$\Delta T_{max} < RT_m2/E_a \qquad (3\text{-}14)$$

where T_m is the initial coolant temperature and the ΔT_{max} is measured at a specific height.

Equation (3-14) is similar to Equation (3-12) for the well-agitated reactor, where the ΔT_{max} is given as RT^2/E_a in which T is the temperature of the stirred reaction mass. A version of Equation (3-14) which applies to unstirred liquids without convection is:

$$\Delta T_{max} = \Theta R(T_m)^2/E_a \qquad (3\text{-}15)$$

in which T_m is the coolant temperature and Θ is the shape factor (slab = 1.19; cylinder = 1.39; sphere = 1.61).

The calculated ΔT_{max} for most reactions is not very large. For a reaction at 100°C that doubles in rate with an increase of 10°C, the value of ΔT_{max} is only 14°C.

To get an idea of the possible effects of a runaway, it is useful to calculate or to determine experimentally the adiabatic temperature rise, and to consider the effect of this temperature increase on the system. An adiabatic temperature rise of 150°C or above is considered a strongly exothermic situation that could result in loss of containment.

3.2.3.2 Heat Transfer in Agitated Vessels

The heat generation varies during processing if an exothermic reaction is carried out in a batch reactor. The temperature can be controlled by external

3.2 REACTOR, HEAT AND MASS BALANCE CONSIDERATIONS

cooling. However, care must be taken to assure that the temperature of the coolant (T_m) is not too low. For example, in the case of a semi-batch reactor, a low T_m could cause a decrease in the reaction mixture temperature resulting in lower reaction rates. This will lead to accumulation of reactants with consequent temperature runaway [175, 178] as shown previously in Figure 3.10.

The factors that can affect the rate of heat transfer within a reactor are the speed and type of agitation, the type of heat transfer surface (coil or jacket), the nature of the reaction fluids (Newtonian or non-Newtonian), and the geometry of the vessel. Baffles are essential in agitated batch or semi-batch reactors to increase turbulence which affects the heat transfer rate as well as the reaction rates. For Reynolds numbers less than 1000, the presence of baffles may increase the heat transfer rate up to 35% [180].

To provide a heat transfer surface in a batch, semi-batch, or continuous-stirred reactor, a jacket, an internal coil, or both can be used. Coils are less expensive than jackets, they provide for higher heat transfer film coefficients, can permit higher internal operating pressures, and are easier to maintain. On the other hand, for example, if a highly viscous material is being processed with an large agitator designed to be close to the vessel wall, a coil cannot be used and a jacket is needed. A possible cause for a runaway reaction is the presence of contaminants. If a minimal cross-contamination between batches is required, a jacket is preferred to permit easier cleaning of the reactor. Thus, from the viewpoint of safety, a jacket may be appropriate. The jacket heat transfer area, however, is limited by the vessel geometry.

Another technique of cooling is to use a boiling solvent to withdraw heat from the reaction mass, usually in a reflux mode as has been discussed previously. An adequate supply of solvent and the appropriate coolant supply to he condenser must obviously be assured.

The choice of the heat exchange system also depends on the overall purpose of the reactor.

The heat transferred in an agitated vessel can be expressed by the basic standard equation:

$$q = UA_s \Delta T_{lm} \tag{3-16}$$

where ΔT_{lm} is the mean logarithmic temperature difference.

For jackets, typical values for the overall heat transfer coefficient U lie between 100 and 600 J/(m^2)(sec)(°C) for cooling and between 200 and 1000 J/(m^2)(sec)(°C) for heating duty. For coils, these values are respectively 200 to 800 J/(m^2)(sec)(°C) and 600 to 1500 J/(m^2)(sec)(°C).

It is important to calculate U accurately to determine the required heat transfer area for a reactor. Typical expressions to calculate overall heat transfer coefficients for agitated vessels are presented in [174, 180] and generally in standard chemical engineering texts and reference books.

3.3 ACQUISITION AND USE OF PROCESS DESIGN DATA

3.3.1 Introduction

The discussions in Sections 3.1 and 3.2 show that the interaction among enthalpies of reaction, reaction kinetics, and surrounding conditions is of paramount importance relative to the existence of potential thermal hazards such as runaways. Whereas valuable information on parameter sensitivity can be estimated by a theoretical approach, it remains of vital importance to evaluate hazards by appropriate and adequate laboratory tests to obtain information on the rates of heat and gas generation, and the maximum quantities of heat and gas involved. Materials which are real to the process should be used in tests to assure that the effects of any contaminants are recognized.

Two sources to obtain this necessary information are the use of data bases and through experimental determinations. Enthalpies of reaction, for example, can be estimated by computer programs such as CHETAH [26, 27] as outlined in Chapter 2. The required cooling capacity for the desired reactor can depend on the reactant addition rate. The effect of the addition rate can be calculated by using models assuming different reaction orders and reaction rates. However, in practice, reactions do not generally follow the optimum route, which makes experimental verification of data and the determination of potential constraints necessary.

Test equipment that can be used for this purpose and typical test results are described in this section. A general strategy to be adapted for a particular reaction system is defined. The parameters to be identified and the sensitivity of the system to the parameter variations must be understood. A test strategy may be relatively simple and comprise only a review of temperature variations to determine the activation energy and the Arrhenius constant factor. However, a strategy may also become complicated and include many studies, for example, into the influence on the heat generation of impurities, charging rates, and catalysts. At an early stage of hazard assessment (during the exploratory stage), an inherent safety approach should be considered for process design as discussed in Section 3.1.1. Larger scale equipment may be required to obtain data concerning the hazard assessment for scale-up [182, 183]. In many cases, though, it is absolutely necessary to obtain data regarding heat generation prior to pilot plant trials.

3.3.2 Bench-Scale Equipment for Batch/Tank Reactors

Since the early 1970s, studies on small-scale batch and semi-batch reactors have been carried out in the industrialized community. These studies resulted, in some cases, in the development of commercially available bench-scale reactors (BSR), such as the RC1 from Mettler-Toledo [184], and the Contalab

3.3 ACQUISITION AND USE OF PROCESS DESIGN DATA

from Contraves [185]; the latter has been bought out by Mettler-Toledo and its availability is now being reduced. These and similar types of equipment enable investigations on parameter sensitivity and optimization studies, and may lead to reduction of certain pilot plant activities.

In the late 1980s, more specialized bench-scale equipment was developed, such as the polymerization reactor [186], and a unit for catalytic reaction studies [187].

Techniques such as adiabatic calorimetry (Dewar calorimetry) were by then well established [2, 118, 119]. All these techniques can be used for obtaining data to design for the prevention of runaway reactions, that is, to design for inherent plant safety.

In the United States, the development in the late 1970s was more focused on venting, especially for two-phase flow venting (DIERS methodology). This work resulted in the development of the Vent Size Package (VSP) and, more recently, the Reactive System Screening Tool (RSST), a more user-friendly version of the VSP; these two units are available from Fauske & Associates. In Europe, the PHI-TEC equipment, similar to the VSP, is marketed. With this type of equipment, information on runaway behavior can be obtained through the proper experimentation along with the general vent sizing data.

Small scale equipment used for reactor process development in the biochemical field, such as small scale fermentation reactors and closed loop reactors, have their merits for general reactor development. These reactors may be as large as 20 liters in volume, and are well equipped and instrumented. They are not reviewed, however, in this book.

Three different principles govern the design of bench-scale calorimetric units: heat flow, heat balance, and power consumption. The RC1 [184], for example, is based on the heat-flow principle, by measuring the temperature difference between the reaction mixture and the heat transfer fluid in the reactor jacket. In order to determine the heat release rate, the heat transfer coefficient and area must be known. The Contalab [185], as originally marketed by Contraves, is based on the heat balance principle, by measuring the difference between the temperature of the heat transfer fluid at the jacket inlet and the outlet. Knowledge of the characteristics of the heat transfer fluid, such as mass flow rates and the specific heat, is required. ThermoMetric instruments, such as the CPA [188], are designed on the power compensation principle (i.e., the supply or removal of heat to or from the reactor vessel to maintain reactor contents at a prescribed temperature is measured).

3.3.2.1 Reaction Calorimeter (RC1)

The RC1 Reaction Calorimeter is marketed by Mettler-Toledo. The heat-flow calorimetric principle used by the RC1 relies on continuous measurement of the temperature difference between the reactor contents and the heat transfer fluid in the reactor jacket. The heat transfer coefficient is obtained through calibration, using known energy input to the reactor contents. The heat transfer areas is obtained by physical measurement of the virtual volume of the reactor contents. The heat generated in an exothermic reaction is transferred into the reactor jacket. Heat transfer fluid circulated in the jacket is controlled dynamically to maintain the reactor contents at a prescribed temperature through the mixing of material from hot and cold reservoirs in the appropriate quantities. Any change in the reactor jacket temperature is achieved by adjusting the respective flow rates from the hot and cold reservoirs. As a result of the very high heat transfer fluid pumping rates, uniform temperature is ensured throughout the reactor jacket.

The RC1 reactor system temperature control can be operated in three different modes: isothermal (temperature of the reactor contents is constant), isoperibolic (temperature of the jacket is constant), or adiabatic (reactor contents temperature equals the jacket temperature). Critical operational parameters can then be evaluated under conditions comparable to those used in practice on a large scale, and relationships can be made relative to enthalpies of reaction, reaction rate constants, product purity, and physical properties. Such information is meaningful provided effective heat transfer exists. The heat generation rate, q_r, resulting from the chemical reactions and/or physical characteristic changes of the reactor contents, is obtained from the transferred and accumulated heats as represented by Equation (3-17):

$$q_r = q_t^* + q_a \tag{3-17}$$

in which q_t^* is the exchanged heat rate $UA_s(T_r - T_j^*)$, and q_a is the heat accumulation rate in the reactor: $mC_p(dT/dt)$. Additional factors such as the sensible heat loss of the addition: $q_d = m_d C_{p,d}(T_r - T_d)$, heat loss by reflux: $q_f = m_f C_{p,f}(T_r - T_f) \times m_f \Delta H_v$, and other heat effects can be introduced into Equation (3-17). The subscripts here are relatively obvious from the text description. In the case of reflux conditions, the equation is correct only if all evaporating components reflux completely (i.e., without loss of mass). This can, of course, be verified by measurements.

The total heat effect is obtained by integration of q_r with respect to reaction time:

$$q = \int_0^t (q_r \times dt) \tag{3-18}$$

For single reactions, conversion at any point in time is proportional to the extent of the heat generation. Checking by analytical methods can be conducted to verify the heat production/conversion relationship. Conversion can also be checked by an on-line balance which is possible with well instrumented systems.

The RC1 is designed to simulate closely the operation of large scale batch and semi-batch reactor systems. The RC1 equipment consists of the reaction vessel, overhead condenser for reflux/distillation operations, receiver, metering pumps, and a heat transfer fluid heating/cooling circulating unit. Me-

3.3 ACQUISITION AND USE OF PROCESS DESIGN DATA

chanical stirring is provided, and the rate of stirring can be adjusted to meet the requirements of the reaction under investigation. All aspects of RC1 operation (i.e., reactor contents and jacket temperatures, reactant addition rates, stirring rates, and safety interlocks) are controlled by a microprocessor. All critical process variables are recorded by the microprocessor during the course of the experimental run. A schematic diagram of the RC1 is shown in Figure 3.11.

Reactor vessels range in volume from 1.2 to 2.5 liters. The size used may be dependent on the pressure application requirements. A high-pressure reactor can run up to about 60 bar. Porting is provided on the reactor vessel for auxiliary probes (e.g, oxygen sensors, pH measurement, and on-line analysis). These probes may be inserted directly into the reactor contents. Porting is also available for the controlled addition of gas, liquid, or solid materials.

Advantages/Disadvantages of the RC1
The RC1 is an automated laboratory batch/semi-batch reactor for calorimetric studies which has proven precision. The calorimetric principle used and the physical design of the system are sound. The application of the RC1 extends from process safety assessments including calorimetric measurements, to chemical research, to process development, and to optimization. The ability of the RC1 to generate accurate and reproducible data under simulated plant scale operating conditions may result in considerably reduced testing time and fewer small scale pilot plant runs.

Heat may be lost due to condensation at the reactor vessel head, but this can be minimized through appropriate insulation.

Although reactions can be run using various parameters, care must be taken to understand fully the impact of such operational changes on the potential for thermal runaway in the RC1 system. Appropriate safeguards should be used to ensure that a thermal runaway, if initiated, can be controlled so as to avoid injury to laboratory personnel and damage to equipment.

3.3.2.2 Contalab
The Contalab, initially supplied by Contraves, was purchased by Mettler-Toledo, which is now placing less emphasis on this design than on the RC1. Some comments here are appropriate, however, since it is another type of bench-scale calorimeter, and units continue to be used. Its measuring system is based on the heat balance principle, in which a heat balance is applied over the cooling/heating medium. For this purpose, both the flow rate of the coolant and its inlet and outlet temperatures must be known accurately. Figure 3.12 is a schematic plan of the Contalab.

The simplified heat balance is described by the following equation:

$$q = mC_p(dT/dt) + \psi_m C_{p,m}(T_{m,i} - T_{m,o}) + \psi_{ref}C_{p,\,ref}(T_{ref,\,i} - T_{ref,\,o}) \quad (3\text{-}19)$$

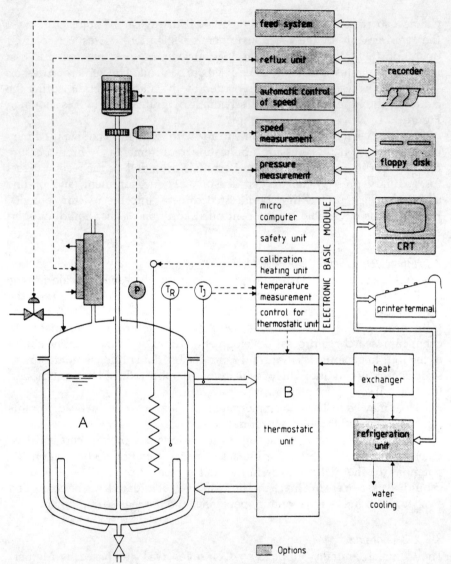

FIGURE 3.11. Modular Design of the Bench-Scale Reactor (RC1).

in which ψ is the mass flow rate and the subscripts are relatively obvious from the text and from Figure 3.12.

The major advantage of this type of calorimeter is that the heat balance principle can easily be applied to the reflux condenser as well, which enables a simpler investigation of processes under reflux conditions. Another advantage is its independence of the heat transfer coefficient at the reactor wall.

3.3 ACQUISITION AND USE OF PROCESS DESIGN DATA

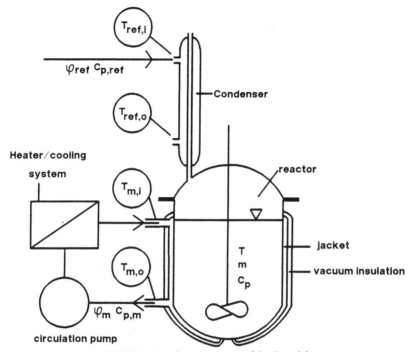

FIGURE 3.12. Schematic Design of the Contalab.

There are some disadvantages, however. The measuring principle reduces the response time when changes in the reaction temperature occur. Therefore, the possibility that temperature gradients exist in the reaction mixture is more significant than with the RC1. Both principles of calorimetric operation are quite practical. though [189].

The usable volume of the Contalab reactor is 1 liter; the operating pressure ranges from vacuum to 1.5 bar; the operating temperature ranges from −30 to 200°C, the heating/cooling capacity is about 1 kW, and the smallest measurable heat flux is 1.5 kcal/hour.

3.3.2.3 CPA ThermoMetric Instruments

The CPA [188], marketed by ThermoMetric AB (Sweden), is frequently used in Europe. It operates on the principle of power compensation, which is based on the supply or withdrawal of heat to and from the reactor, respectively, in order to keep the temperature at the set-point and, thus, to compensate for energy differences (either shortage or surplus). The heat is supplied or withdrawn by means of special (Peltier) elements, which produce a cold or a hot surface area if subjected to an electrical current. An accurate measurement of the heat supply/withdrawal is possible as the heat flow is directly proportional to the current supplied to the Peltier elements.

FIGURE 3.13. The CPA System

The CPA reactor is a cylindrical, double-walled, glass vessel as shown in Figure 3.13. The bottom and top of the reactor are made of steel. During operation, the bottom and top are located in a fixed position in the stainless steel thermostat which also acts as a safety shield. The usable volume of the reactor is 40 to 180 cm^3. Operating pressures range from vacuum to 20 bar, and operating temperatures range from −50 to 150°C. Ports are provided in the top and base plates for auxiliary probes, such as pH or oxygen sensors, which can be inserted directly into the reactor contents. Different types of mixers can be used, including a torque transducer for continuously measuring the mixing power input.

3.3.2.4 Quantitative Reaction Calorimeter

The quantitative reaction calorimeter is described in [190]. The purpose of this instrument is the evaluation of thermodynamic and kinetic properties of chemical reactions. The calorimeter has provisions for either isothemal or temperature programmed modes of operation over a wide range of temperatures, with simultaneous or sequential addition of several reactants, and variable-rate agitation. The metal reaction vessel can be removed from the main housing of the calorimeter, enabling the use of vessels made form various materials of construction for optimum corrosion resistance. The capacity of the vessel is at least 100 cm^3 to provide good control of reactant additions and agitation. Heat is exchanged through the bottom of the vessel. The observed detection threshold is 50 mW, and a heat flux of several hundred watts can be dissipated. Provision is possible for the measurement of other parameters such as pH.

The reaction vessel is provided with a pressure relief venting system which includes a small spring-loaded relief valve; the design also permits the lifting of the entire cover if necessary.

A sketch of the quantitative reaction calorimeter is shown in Figure 3.14.

3.3.2.5 Specialized Reactors

In the literature, examples are given of bench-scale equipment designed for special fields such as for polymerizations [186] and for kinetic studies of catalytic reactions [187].

The polymerization reactor is of the heat-balance type because of the change in the heat transfer characteristics of the reaction mass during the polymerization. As the viscosity increases, the rate of heat dissipation by mixing will generally decline, which must be taken into consideration in setting up the equipment and in taking the appropriate measurements.

The bench-scale unit for the study of catalytic reactions has been designed with features such as accessibility, isothermal operation, and catalyst pretreatment. The use for catalytic screening tests makes easy accessibility a necessity, while the study of kinetics prescribes isothermal operation.

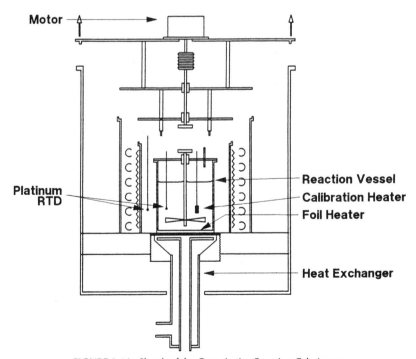

FIGURE 3.14. Sketch of the Quantitative Reaction Calorimeter.

The advantage of such specially designed calorimetric reactors is that optimal results can be obtained for the designated reaction systems. Use for other types of systems, however, is limited.

3.3.2.6 Vent Size Package (VSP)

The original Vent Size Package (VSP) became available first in the United States, and is currently distributed by Fauske and Associates. A very similar type of equipment, called PHI-TEC, is available in Europe.

The primary application of the VSP is to obtain data necessary to calculate the vent design (size and relief setting) for emergency venting of nonvolatile and reactive runaway systems. The calculated vent design is to limit the maximum pressures at the point of emergency venting to acceptable levels. A secondary application is to provide thermal stability data for reactive systems.

The data required for the emergency vent design includes [191]: (1) the thermokinetic and pressure history monitored under near adiabatic conditions, (2) the character of the type of vented system (vapor, gassy, or hybrid), (3) the phase of the vented material (vapor, liquid, or two-phase), and (4) the degree of two-phase disengagement (turbulent, bubbly, or homogeneous). To determine these characteristics, the VSP defines the system as viscous (100 cp) or nonviscous, and also whether or not it has a foaming tendency.

The VSP, shown in Figure 3.15, consists of a 120 cm^3 stirred metal test cell (7 mm in thickness) which is heated on its outer surface with the main heater. There is also an auxiliary heater wired on the outer surface of an aluminum can which is separated from the test cell by a uniform layer (4 mm thick) of insulation. The latter acts as an adiabatic guard heater to minimize heat losses. The relatively low pressure rating of the test cell is compensated by a pressure control system which keeps the pressure outside the test cell equal to the pressure inside the cell. This enables the application of thin-walled cells with low thermal inertia even for reactions which cause a rapid pressure increase. The VSP is characterized by a phi-factor as low as 1.05 when relatively large samples are tested. This is a major advantage over, for example, the ARC which has a phi-factor between 1.5 and 6 depending on the test cell used. Proper use requires careful determination of test conditions and application of the corresponding DIERS technology.

A disadvantage is that the maximum sample size is only 100 cm^3 which may not, then, be fully representative. Extrapolation of test results to plant facilities must therefore be carried out with good judgment [193].

The pressure control system can equilibrate pressure change rates up to 20 bar/s. The outside guard heater can cope with temperature changes up to 100°C/min. A heat loss rate of less than 0.1°C/min can usually be achieved below 350°C and 25 bar [194]. The tests are run with either open or closed test cells in a closed outer bomb, which can be vented under specified conditions. The containment vessel can withstand pressures up to 100 bar. The use of a closed test cell results in the most severe pressure and temperature changes.

3.3 ACQUISITION AND USE OF PROCESS DESIGN DATA

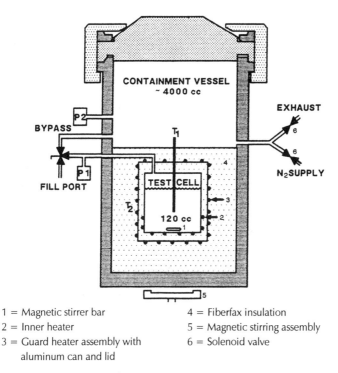

1 = Magnetic stirrer bar
2 = Inner heater
3 = Guard heater assembly with aluminum can and lid
4 = Fiberfax insulation
5 = Magnetic stirring assembly
6 = Solenoid valve

FIGURE 3.15. Vent Size Package (VSP) Test Cell.

By using open or vented cells, data are obtained to determine if the full-scale reactor vent flow will be single or two-phase, and if a tempered or nontempered reaction occurs [192]. A tempered reaction is one that can be controlled through the latent enthalpy of evaporation.

Another feature of the VSP is the ability to differentiate between foamy and nonfoamy behavior of the system under investigation, which is especially important in considering venting requirements. The following types of foaming systems are recognized:

- *Vapor systems*—the pressure generated by the runaway reaction is due to increasing vapor pressure of the reactants, products, and/or solvents as the temperature rises,
- *Gassy or nontempered systems*—the total pressure is due to the fact that the reaction produces permanent gases, and
- *Hybrid systems*—the total pressure is due to both an increase in vapor pressure and in permanent gas production.

With the assumption of two-phase turbulent flow, a simplified method has been developed [195] for estimating emergency vent sizes which is discussed in Section 3.3.4.7.

The VSP is a valuable tool in assessing temperature and pressure data with a relatively small sample in a short period of time. The low thermal inertia cells represent an advantage. A disadvantage is the relatively small contents of the test cell, which requires a good representative sample. Also, there is complexity in applying results to the DIERS technology.

Some examples, such as thermal polymerization of styrene and decomposition of di-t-butyl peroxide, are given in [194], both treated as first-order reactions. The activation energy found for the decomposition of di-t-butyl peroxide agrees well with the literature value. From the pressure data, it appears that the initial pressure rise is caused by the evaporation of toluene, present as a solvent. At higher temperatures, the gases generated by decomposition are the main contributors to the pressure rise.

3.3.2.7 Reactive System Screening Tool (RSST)

The Reactive System Screening Tool (RSST), marketed by Fauske and Associates, is a relatively new type of apparatus for process hazard calorimetry [192, 196–198]. The equipment is designed to determine the potential for runaway reactions and to determine the (quasi) adiabatic rates of temperature and pressure rise during a runaway as a function of the process, vessel, and other parameters.

The basic features of the RSST are illustrated in Figure 3.16. The RSST consists of a small open spherical glass test cell with an inner volume of about 10 cm^3. The test cell is placed in an outer stainless steel containment vessel (inner volume about 500 cm^3) which serves both as a pressure containment vessel as well as a means to provide back pressure simulating a plant unit. The containment vessel can withstand pressures up to 35 bar.

The temperature of the spherical test cell is recorded by a thermocouple and the pressure of the containment vessel is measured by a pressure transducer. The test cell is covered with insulation. It is placed in a stainless steel unit with a lid to reduce the heat loss to the surroundings. A single heater element is located in the center of the test cell to compensate for heat loss and to initiate a runaway by heating the sample. The immersed heater is replaced by an externally wound heater assembly at the base of the glass test cell when solid systems are investigated. Via feedback control, sufficient power is introduced to permit a linear temperature increase rate such that heat losses are properly balanced. In case of a reactive substance or system, . this linear increase in heating is continued during the runaway reaction and is thus added to the reaction energy release (over-adiabatic mode), which implies that the effect of heat loss on energy generated during the runaway is reduced. The RSST provides the capability of running experiments at selected set pressures.

The RSST apparatus can be operated with its own controller or with a computer interface. Heating rates depend on the C_p of the reactive sample, and can be varied from 0.25°C/min to approximately 2°C/min. Isothermal experiments can also be run. The contents of the test cell can be mixed with a

3.3 ACQUISITION AND USE OF PROCESS DESIGN DATA

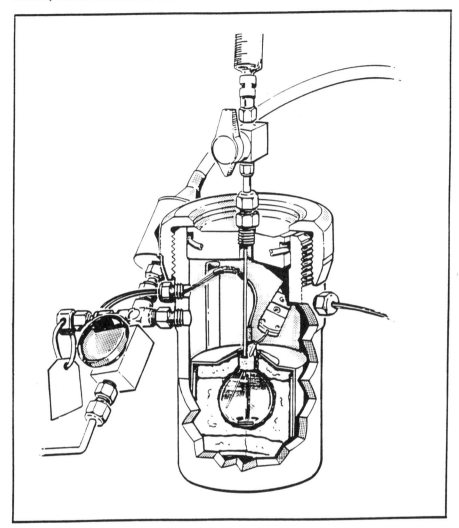

FIGURE 3.16. Schematic of the RSST Showing the Glass Test Cell and the Containment Vessel.

magnetic stirrer. Reactants can be added to the contents of the test cell during an experiment through a feed tube.

A typical temperature–time curve from a run in the RSST is shown in Figure 3.17.

Assessment of Results
Data acquired from an RSST experiment show the potential of a runaway reaction (reactive or nonreactive), the temperature history of the runaway, and the rates of temperature and pressure rise (the latter in the case of gas-produc-

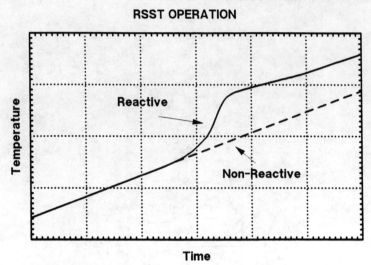

FIGURE 3.17. Typical Temperature–Time Curve of an RSST Experiment.

ing reactions). This permits the determination of the energy release and gas release rates. Calculations of the energy release are, in principle, the same as discussed in Chapter 2 for the ARC, taking into account the quasi-adiabatic conditions. These data can be combined with simplified methods [192, 197] to assess vent and safety requirements for reactor and storage systems. A comprehensive comparison of results obtained in RSST and in VSP equipment was published [192].

A comparison of RSST and VSP characteristics is shown in Table 3.3.

Advantages/Disadvantages: RSST
Advantages of the RSST are its relatively low cost and its availability to permit a quick evaluation for potential runaways. Pressurized conditions may be used. The temperature-time curve and the concurrent pressure increase (of the containment vessel) can be recorded, which are measures of the reactivity hazards of the substance or reaction under investigation. The temperature-time curve shows the lowest temperature at which a runaway can be detected in the test system (initiation temperature, T_0).

The primary disadvantage of the equipment is that it cannot run a closed system test. Loss of material could impact on the initiation temperature and the magnitude of the decomposition. The loss of material, further, can lead to an underestimation of the necessary vent size, although the material loss can be significantly reduced by increasing the test back pressure.

3.3 ACQUISITION AND USE OF PROCESS DESIGN DATA

TABLE 3.3
Characteristics of the RSST and VSP

Test Equipment	Typical Mass of Sample (g)	Typical Sensitivity (W/kg)	Thermal Inertia of System	Possible Data Acquired
RSST	5 to 10	200	<1.1	T_0, ΔH_O, ΔH_d, ΔH_r, dp/dt, dT/dt, gas release, pressure data
VSP	10 to 100	100	1.05	T_0, ΔH_O, ΔH_d, ΔH_r, dp/dt, dT/dt, gas release, pressure data

The RSST cannot be used with systems having significant resistance to effective stirring, for example, viscous liquids.

The test is primarily a screening tool relative to reactivity of substances and reaction mixtures and is highly useful for that purpose. The determined initiation temperature is approximate. The energy calculations based on temperature increase and heat capacities are semi-quantitative because of the quasi-adiabatic mode of the system operation. The method of insulating the test cell results in moderate reproducibility of temperature rise and related pressure rise. Another disadvantage is the relatively small sample quantity with respect to full scale quantities; thus, there could be a problem in that the sample may not be truly representative.

It should be emphasized that the results from RSST experiments use simplified calculation methods and give estimated values for stability and vent sizing. For large scale purposes, more dedicated and accurate measuring techniques may be needed for further hazard evaluation determinations. Due to the relatively recent development of this apparatus, comparison with other stability test methods is not yet comprehensive.

3.3.3 Process Safety for Reactive Systems

This section discusses how the interpretation, evaluation, and correlation of test results from bench-scale equipment can be integrated into an approach to inherent process safety involving reactive systems.

3.3.3.1 Test Plan
The frst step in the approach is to draft a test plan which will generate both the required technical data and the sensitivity of the reaction parameters. If large scale continuous processing of very energetic materials is considered,

testing can still be performed in a semi-batch reactor because the cooling capacity of the bench-scale equipment is considerably greater on a unit basis than the capacity of the large scale equipment.

Table 3.4 lists the essential questions on safety aspects, the information requirements, and the selected methods available to obtain the desired information and data.

The answers to any of these questions can be incorrect because of poor assumptions. For example, accumulation of reactants or intermediates may be caused by use of the incorrect kinetic assumptions, too rapid a feed rate, too low a reaction temperature, incorrect reaction initiation, insufficient mixing, and by impurities. In the same way, several causes can be given for a higher heat generation than originally estimated.

TABLE 3.4
Essential Questions on Safety Aspects of Reactions

Question	Data Required	Selected Methods of Investigation
1. What is the potential temperature rise by the desired reaction? What is the rate of the temperature rise? What are the consequences? What is the maximum pressure?	• Enthalpy of desired reaction • Specific heat • Vapor pressure of solvent as a $f(T)$ • Gas evolution	• Tables of data • Thermodynamic data • Calculations; estimations • DTA/DSC • Dewar flask experiments • Reaction calorimetry with pressure vessel • Thermometry/manometry • ARC/RSST/VSP
2. What is the potential temperature rise by undesired reactions, such as from contaminants, impurities, etc.? What are the consequences? What is the maximum pressure?	• Data from No. 1 • Enthalpy of undesired reaction • Specific heat • Rate of undesired reaction as a $f(T)$	• See No. 1 • DTA/DSC • Dewar flask experiments • ARC/RSST/VSP
3. Is reactant accumulation possible? What are the consequences?	• Steady state concentrations • Kinetic data • Data from 1 and 2	• Reaction calorimetry combined with analysis • Potential energy by DSC/DTA • RSST/VSP
4. What is the temperature rise due to physical aspects of the system?	• Heat transfer data • Agitation (power input) • Pumps (power input) • Radiation	• Design data

3.3 ACQUISITION AND USE OF PROCESS DESIGN DATA

Undesired reaction products must be considered in the safe handling of materials and in writing procedures. Cooling capacity, reactant addition, reflux conditions, effect of catalysts, impurities, contamination, and sensitivity to solvents can be studied showing the effect on heat generation, yield, and selectivity. Although test plans have a similar general approach, they can be very diverse and are dependent on the triangle characteristics discussed in Chapter 1 (Figure 1.1).

3.3.3.2 System under Investigation

The physical states of the components in the system under investigation have a significant impact on the potential hazards involved and thus represent an important aspect of the evaluation. Significant research has been conducted on liquid–solid (solid catalysts), liquid-liquid (separation of products) and liquid–gas (aeration, oxidation) systems [199].

The type of agitator, its location, the application of one or more multiple mixing planes, the mixing rate, vessel geometry, and baffling are important factors in a mechanically agitated reactor system which influence yield and selectivity. In particular, for example, special attention should be given to agitation of two-phase systems.

A turbine type agitator is commonly used for liquid–solid systems. Mixing rates depend on the forces required to suspend all solid particles. Minimum levels can be determined for: (1) lifting the particles, and (2) for suspending them in an homogeneous manner [200]. Similar requirements apply to liquid–liquid systems. For cases where two poorly miscible fluids of about equal volume are used in the reaction, the mixer is placed at the interface. For a bench-scale experimental system of about 2 liters capacity, the minimum rotational speed to obtain well-dispersed system is 300 to 400 rpm [201], depending on the type of mixer. This rotational value decreases as the vessel volume increases.

For gas–liquid systems at low mixer speeds, the gas may flow through the reaction liquid resulting in a small interfacial area. At higher mixing rates, the gas bubbles decrease in size, thus enlarging the interfacial area. An increase in gas flow (larger superficial velocity or gas load) may ultimately lead to flooding the reactor [202].

Scale-up rules have been established for liquid–liquid, liquid–solid, and liquid–gas systems [199].

The state of mixing determines the mass transfer, especially for heterogeneous systems. The mass transfer in a Liquid–liquid system can take place only at the interface of the two layers. An increase in interfacial area in the event of a sudden increase in mixing rate, for example, at a start-up following a stoppage of the agitator, will lead to a rapid increase in conversion rates and hence in heat production.

For liquid–gas reactions normally run at low agitator speeds, increasing the agitator speed, whether intentionally or inadvertently, can lead to higher mass transfer between the phases and thus result in higher reaction rates. The possibility of mixing rates occurring other than the normal design operation, with the subsequent effects on reaction kinetics and potential heat generation, must be considered in the design of the system.

In case of insufficient mixing or a poorly dispersed system, accumulation of reactants may occur. This is often the case at the start of a process. If the conversion then gains momentum, a runaway may occur because of this accumulation.

It is obvious that the effect of mixing must be taken into account in order to get reliable kinetic data. Such factors as $k_g a$ and $k_l a$, the mass transfer coefficients, are largely determined by mixing efficiency, especially in heterogeneous systems.

3.3.3.3 Test Results

Reaction calorimetry provides information on the maximum heat generation at process temperatures and on the adiabatic temperature rise. This ΔT_{ad} provides insight into the worst-case temperature consequences.

In order to determine the thermokinetics in experimental runs, the temperature is varied between certain limits depending on the thermal stability of the original components, the intermediates, final products, and by-products. For example, if the optimum temperature is T_{opt}, the temperature may be varied between $T_{opt} \pm 25°C$.

For a simple batch reaction, the activation energy and the pre-exponential factor are determined directly from an Arrhenius plot. In a semi-batch system, the reaction kinetics have to be known because the concentrations of reactants differ at the conditions of maximum heat generation for various reaction temperatures (and a constant feed rate). This makes it difficult to obtain the activation energy for a semi-batch system. The problem can be overcome by determining an apparent activation energy or by fitting different reaction orders to obtain a value for the activation energy. Reference may be made to Figure 3.18 [167].

The total heat effect is obtained by similar experiments and calculations using Equation (3-18) discussed previously in Section 3.3.2.1. The total heat effect is the integral of q_r over the reaction time. The conversion at any time during the reaction can be estimated from the ratio of the integral of q_r to any time t over the total integral. Typically, chemical analyses show good agreement between calculated and chemically determined values.

The adiabatic temperature rise can be calculated from the total heat effect and the specific heat of the reactor contents. Both parameters can, for example, be determined by using appropriate procedures for the Mettler-Toledo RC1 or Contalab. The adiabatic temperature is calculated by:

$$\Delta T_{ad} = \frac{Q}{mC_p} \tag{3-20}$$

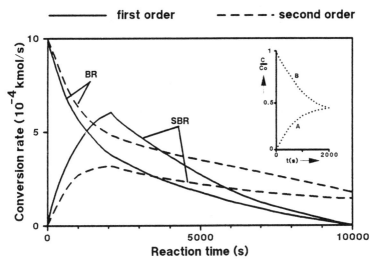

FIGURE 3.18. Semi-Batch versus Batch Operations for First- and Second-Order Kinetics.

For small vessels and slow reactions, corrections must be made because of the heat content of the reaction vessel itself. For large-scale reaction vessels and for rapid reactions, the system will be close to adiabatic operations. This aspect must be taken into account in scale-up. In effect, the extrapolation of data obtained in small-scale equipment has limitations as discussed in [193]. In case of a runaway, the maximum temperature in the reaction system is obtained from the adiabatic temperature rise, that is, $T_{max} = (T_r + \Delta T_{ad})$. In reality, the adiabatic temperature rise is significantly underestimated if other exothermic reaction mechanisms occur between T_r and $(T_r + \Delta T_{ad})$. Therefore, a determination must be made to see if other exothermic events, which may introduce additional hazards during a runaway, occur in the higher temperature range. This can determine if a "safe operating envelope" exists.

In most bench-scale reaction instruments, it is also possible to perform adiabatic experiments, although precautions have to be taken to avoid an uncontrollable runaway in the final stages. From these types of experiments, the temperature constraints at which, for example, side reactions or decomposition reactions start, together with the possible control requirements, can be obtained. If the adiabatic temperature rise may exceed, say, 50 to 100°C, it is safer to use other methods to obtain similar information, such as the DSC, ARC, or Sikarex, because these instruments use relatively small amounts, thereby decreasing the potential hazard of an uncontrollable runaway event in the test equipment.

In general, all heat that is generated during a reaction in a bench-scale experiment is exchanged with the cooling medium. The maximum heat generation and rate determine the maximum required cooling capacity; thus, the type and capacity of the cooling system, from air cooling to highly turbulent liquid cooling, can be defined. If the cooling capacity is limited, the maximum heat generation rate has to be decreased, for example, by decreasing the reaction temperature in a batch reaction and decreasing the addition rate in a semi-batch reaction. Use of too low a coolant temperature in batch reactions, as another possible corrective action, can lead to a runaway as described previously in Chapter 3.

3.3.3.4 Malfunction and Process Deviation Testing

A part of the test plan must include testing for the consequences of equipment malfunction, deviations in process conditions, and human error. Bench-scale equipment, for example, the RC1, is quite suitable for such experiments. By analysis of the process, critical conditions can be defined, which then need to be tested in order to be able to proceed safely from the laboratory to pilot plant studies. In testing abnormal conditions or process deviations, caution is required to assure that no uncontrollable hazard is created in the laboratory. Typical deviations, including impact on the process, are discussed in the following paragraph.

Agitator failure effects
Agitation failure during a reaction can affect the following system parameters:

- total heat generation and heat generation rate,
- mass transfer,
- cooling,
- accumulation of reactants, and
- kinetic data determination (testing).

Not all of the above are mutually independent.

The state of mixing generally controls the mass transfer. In a liquid–liquid system, for example, the reaction rate is based on the mass transfer which depends on the interface area of the two liquid layers. This area is dramatically changed by a change in the mixing rate. If, for example, the agitator is started late, the increase in mass transfer area will lead to a rapid increase in the conversion rate and hence in the heat production rate.

Insufficient mixing may lead to a poorly dispersed system which can result in an accumulation of reactants. The reaction may not start readily or perhaps will start only after a considerable accumulation of reactants leading to a runaway.

If agitation fails during a semi-batch operation, the transfer of heat will essentially stop. The resulting increase in temperature depends on the concentration of the reactants at that moment, the global kinetics, and the mass transfer rate. The effect of the temperature increase is easily simulated in a reaction calorimeter both with and without addition of reactants.

3.3 ACQUISITION AND USE OF PROCESS DESIGN DATA

A reaction at the interface caused by a loss of mixing can eventually lead to a runaway itself. Examples include nitration processes and the well-known Seveso incident where agitation stopped in a reactor during the manufacture of trichlorophenol; this led to higher than normal temperatures and increased production of the undesirable side product 2, 3, 7, 8-tetrachlorodibenzo-*p*-dioxin (commonly referred to as "dioxin"), ultimately resulting in a vapor release to the atmosphere.

The start-up procedure after an agitation failure can be studied in detail in bench-scale testing as well.

Loss of cooling
The effect of cooling is determined by:

- volume of reactor, reaction mass, and associated reaction equipment,
- type of cooling system (e.g., jacket, coil, or external heat exchanger),
- type of coolant and circulation, and
- kinetics/rates.

Either complete or partial loss of cooling will influence the operating temperature, the rates of reaction, and the rates of heat generation. A cooling temperature too low at start-up of the process may lead to an accumulation of reactants, especially if the reaction does not initiate while the charging of recatants continues [203]. This was discussed previously and was illustrated in Figure 3.10 for a semi-batch reactor. Its effect on the energy content of the system depends on the concentration of reactants, the enthalpy of reaction, and on temperature constraints. The effect of cooling loss increases considerably if chain reactions can occur when the temperature increases. Subsequently, the heat release and conversion at the moment of loss of cooling determine the rate at which the runaway occurs and the maximum temperature rise involved.

In practice, large-scale reactors operate close to adiabatic conditions on loss of cooling which causes maximum increases in temperature. In smaller reactors, the temperature increase depends on the heating of coolant and reactor, and the heat loss to the reactor frame and confined coolant as well.

Contamination
Contaminants can be any chemical substance used in the plant that does not belong in the reaction system. This includes coolants, air (oxygen), solvents, lubricants, rust, and so forth. Changes in chemical feed stocks, for example, from using different suppliers, can result in the introduction of contaminants as well. The result of contamination may be:

- lower yield,
- change in kinetics, and/or
- catalysis of the reaction into a runaway.

It is clear that not all possible contaminants can be tested, but sources of contamination must be considered and tests run on the reaction in the presence of the most likely occurring ones. An approach to evaluating the problem of contamination is in the setting-up of a plant material matrix [1]. An example of potential contaminants to be considered, and sometimes overlooked, includes the heat transfer fluids to evaluate the consequences of heat exchanger, coil, or jacket failures. Contaminants which are introduced by other sources, for example, air (oxygen), carbon dioxide, water, metals, lubricants, and greases must also be considered. Also, the effects of chemicals which are used elsewhere in the plant and which could be introduced by mistake should be evaluated and perhaps tested. The possible contaminants in the reactor feeds must also be considered.

The compatibility of likely contaminants with the reactants has to be determined (see Section 2.2.4). Their effect on the reaction or decomposition process, for example, by catalysis, must be known to develop a strategy in case of an emergency upset.

Possible effects of the presence of contaminants may be only a poor yield and not a safety issue as such. However, an increase in the reaction rate, the delay of reaction initiation, or catalysis of parallel reactions may occur, which could lead to a runaway.

When changing the supplier of reactants, the specifications have to be evaluated carefully for the presence of different chemical components as impurities to identify potentially hazardous contaminants.

Autocatalysis (i.e., catalysis by one of the products of the reaction), should also be considered. The impact of the accumulation of the catalytic product must be evaluated, and appropriate consideration given to runaway potential and prevention.

Mischarging

The following items related to mischarging should be considered in any test plan:

- too rapid/too slow charging rates,
- incorrect amounts of reactants charged,
- incorrect reactant concentrations in feed solution,
- wrong order of addition of reactants,
- wrong reactant charged—which substance? what effect?
- charging to remaining heels in vessels.

When the charging rate (semi-batch) or the mass of reactant added in one step (batch) or the continuous feed ratios get out of phase, the heat generation may be larger than expected. This results in difficulties in the transfer of the heat generated, especially in cases where the cooling capacity has a limited safety margin Also, it can lead to accumulation of reactant in a semi-batch operation.

3.3.3.5 Pressure Effect

In general, the pressure of a reaction system can increase for three reasons: (1) evaporation of low boiling chemicals, (2) formation of gaseous by-products as a result of the desired reaction, and (3) production of gases as a consequence of undesired reactions or decompositions. For normal operations, it is imperative to know how deviations in operating conditions affect the gas production. Further, the effect of increased pressure on the reaction rate must be determined to avoid uncontrollable pressure increases in confined systems.

In case of undesired reactions, an emergency may develop. The combination of heat generation and gas production may accelerate the decomposition rate disproportionally.

Gas production and subsequent pressure–time histories can be investigated successfully only in pressure vessels such as the VSP. If the gaseous product dissolves partly in the reaction mixture (i.e., the vapor-liquid equilibrium is changed), careful investigations of the pressure effect within the possible variations of the operating conditions are necessary. Pressurized vessels are also useful to investigate any mass transfer improvement for gas–liquid or gas-dissolved (suspended) solid reactions.

3.3.3.6 Results from the ARC, RSST, and VSP

De Haven [127] gives an overview of the results of accelerating rate calorimeter (ARC) experiments. The ARC was described in Section 2.3.2.3. As mentioned in the previous description, care must be taken in scale-up of results from experiments with relatively high phi-factors. For direct simulation of plant operating conditions, a phi-factor of 1.0 to 1.05 is required. As stated in [127], a decrease in the phi-factor from 2.0 to 1.0 increases the adiabatic temperature rise by a factor of 2, but the maximum self-heat rate increases by a factor of 20. Later in Chapter 3 (Section 3.3.4.6), an example of scale-up of ARC results is given.

The Reactive System Screening Tool (RSST) was described in Section 3.3.2.7. This apparatus is a relatively recent development. Therefore, only limited literature data are available regarding the application of results from this equipment for direct scale-up of reactor systems. The Vent Size Package (VSP) is discussed further in Section 3.3.2.6.

3.3.4 Scale-up and Pilot Plants

This section starts with some general remarks concerning scale-up of chemical reactors. Then the influence of chemical kinetics, heat transfer, and mass transfer on scale-up of reactive systems is discussed. Finally, scale-up from the results of calorimetric equipment, such as the ARC and VSP, is reviewed.

3.3.4.1 General Remarks

Small-scale laboratory tests often are not fully representative of all conditions encountered at the commercial production scale. The following discussion illustrates some of the difficulties of extrapolating results from small-scale testing to a production unit.

The first example involves flammability issues that are not specifically covered in this Guidelines book. However, the discussion here is highly important for safe process design considerations and represents a good example of the problems of scale-up from test data. Runaway reactions may indeed result in the production of flammable gases so an understanding of the scale-up problems is critical.

The lower and upper explosive limits (LEL and UEL) of gases (really gas–air mixtures) are determined in small-scale equipment by spark ignition at ambient temperature and pressure. The flammability range established by this method may deviate from the actual range at the commercial scale. At elevated temperatures, the flammability range is expected to increase. A similar phenomenon occurs at higher pressures, while at lower pressures, ignition may become impossible.

Gas mixtures at the test level are usually homogeneous, whereas the homogeneity of a gas–air mixture may vary considerably on a commercial scale. Thus, there may be pockets of flammable gas–air mixtures interspersed with nonflammable mixtures in the vapor space of commercial equipment. Flows of flammable gas mixtures may not be ignitable at the same conditions under which equivalent static mixtures can ignite. This phenomenon is caused by heat loss by mixing, and by increased heat transfer to the walls of the equipment enclosing the gas. The rate of flow strongly influences the ignitability of flammable gas mixtures because the flow rate directly affects these heat transfer mechanisms. Ignitability also varies with the scale of operation which changes the flow hydrodynamics.

Another example of scale-up effects relates to the storage of chemically unstable substances. Well-established procedures can be followed on a small scale. In a commercial unit, the storage of such materials must be reviewed from the standpoint of critical mass. The heat removal capacity of the equipment must be substantially larger than the spontaneous exothermic rate of heat release in the bulk material. Temperature gradients must also be considered.

These and other scale-up effects must be considered in the appropriate start-up and operation of a commercial unit for which the design and operating procedures are based, at least in part, on experimentation and demonstration on a small scale.

Successful start-up of operations means that production is accomplished safely at planned rates and product is manufactured to the desired quality specifications. Experience shows that, in moving from small scale to commercial equipment, the following variables are important:

- shape (introduces differences in agitation, fluid short-circuiting, or stagnation zones),

3.3 ACQUISITION AND USE OF PROCESS DESIGN DATA

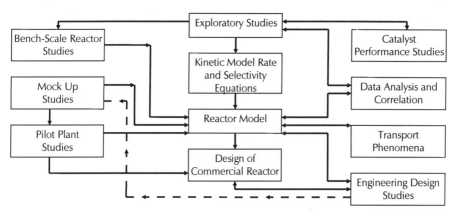

FIGURE 3.19. Typical Structure for Reactor Design.

- mode and scale of operations (results in different residence time distributions),
- surface-to-volume ratios (flow patterns and geometry result in significantly different gradients of concentrations and temperatures),
- materials of construction (may result in different contamination levels),
- flow stability and mixing capacity,
- heat removal,
- wall, edge, and end effects,
- need for storage of intermediates, and
- use of recycle materials, such as solvents.

These components of scale-up manifest themselves through the effects of chemical kinetics, mass transfer, and heat transfer. As an example of the way these factors interrelate to scale-up, the general process of commercial scale reactor design is shown in Figure 3.19, which is similar to presentations in [204, 205].

3.3.4.2 Chemical Kinetics

Table 3.5 shows that the study of chemical kinetics is critical in successful scale-up of catalytic systems, of gas-phase controlled systems, and of continuous tank stirred reactors (CSTR). For scale-up of batch systems consisting of gas or liquid compounds, chemical kinetics and heat transfer effects must be studied because the combination of these phenomenon determine the conditions for a runaway and thus involve the safety of the operation.

Laboratory studies are very important for providing basic knowledge to scale-up of batch reactions. Modeling a batch system is very important as well. When the batch reaction and system are well understood, a large scale-up factor may be applied while still maintaining safe operations.

TABLE 3.5
Reactor Scale-up Characteristics

Phenomena under Study	Gas or Liquid			Gas and Liquid			Catalytic	
	Batch	Tubular	CSTR	Liquid-Phase Control	Gas-Phase Control	Gas Fixed Bed	Gas–Liquid Fixed Bed	Fluidized or Moving Bed
Chemical Kinetics	2	3	1	3	1	1	1	1
Mass Transfer	4	4	4	1	3	4	1	4
Heat Transfer	1	1	2	4	3	3	2	3
Scale-up Methods								
Laboratory studies	1	4	1	3	2	3	2	1
Pilot plants	4	1	3	3	1	1	1	3
Mockups	4	4	3	1	1	4	1	1
Modeling	2	2	1	1	2	2	3	3

1 = critical/very important
2 = necessary/important
3 = desirable/some importance
4 = little value/unimportant

For a gas–liquid reaction which is gas-phase controlling, the chemical kinetics must be well understood. The importance of laboratory studies must therefore be emphasized. However, for successful scale-up, pilot plant studies are very critical because of the difficulties in reliably modeling gas behavior on a small scale (due to hydrodynamics) and its influence on reaction rates.

3.3.4.3 Mass Transfer/Mixing

Agitation plays an important role in process chemistry scale-up. Different mixing characteristics are involved depending on whether the system is liquid–liquid, liquid–gas, liquid–solid, solid–gas, or gas–liquid–solid, and whether it is homogeneous or heterogeneous. Included in the important mixing characteristics for scale-up are minimum agitation rate and the degree of dispersion.

Scale-up calculations are generally performed by the use of dimensionless groups which are ratios of balancing factors. Through a proper use of dimensionless groups, scale-up is possible via similarity through balancing components within the dimensionless group. For example, the temperature

3.3 ACQUISITION AND USE OF PROCESS DESIGN DATA

distribution in a vessel is one feature determined by the dimensionless Biot number, (hx/λ), where h is the film heat transfer coefficient, x is a dimension, and λ is the thermal conductivity. Identical Biot number values in two different vessels result in identical temperature distributions. The Biot number balances heat transfer and thermal conductivity characteristics. If the heat transfer area is increased (i.e., a larger x value), an identical ratio of heat transfer and thermal conductivity is obtained only by either decreasing the heat transfer or increasing the thermal conductivity. Larger temperature gradients will occur if the heat transfer area is increased without corresponding compensation by altering the heat transfer or thermal conductivity of the system. Other numbers, such as the Frank-Kamenetskii number, which is discussed later, are important in scale-up of reactors as well.

Scale-up equations for liquid, gas–liquid, and solid–liquid systems are detailed in [167, 199, 201, 206].

3.3.4.4 Heat Transfer

Factors involved in heat transfer, such as surface-to-volume ratio, agitation characteristics, mixing efficiency, fouling of heat transfer surfaces, scale of operations, and the resulting heat exchanged depend on the system under consideration (e.g., liquid–liquid transfer, liquid–gas transfer, free convection, or forced convection). Standard chemical engineering texts and reference books contain detailed discussions on heat transfer in process equipment. Only a brief summary follows:

Generally, the heat exchanged in reaction vessels can be described by:

$$q = UA_s \Delta T_{lm} \qquad (3\text{-}21)$$

in which the overall heat transfer coefficient U depends on the heat conductivity of the vessel wall(s) and on film coefficients that are based on the flow characteristics on both sides of the wall. This equation is the same as Equation (3-16) discussed in Section 3.2.3.2. For a single wall reactor, the following equation is valid for the general case:

$$\frac{1}{U} = \frac{1}{h_j} + \frac{l}{\lambda} + \frac{1}{h_r}\frac{A_{sj}}{A_{sr}} + FF \qquad (3\text{-}22)$$

where h is the film coefficient, the subscripts "j" and "r" refer to the jacket and reaction side, respectively, l is the wall thickness, λ the thermal conductivity of the wall material, and FF represents the fouling factors.

Heat exchange in stirred reactors is described in [207]. By using dimensional analysis of heat flow and energy balance equations, the Nusselt number, containing h_r, can be expressed as a function of the Reynolds number and the Prandtl number:

$$N_{Nu} = f(N_{Re} \times N_{Pr}) \qquad (3\text{-}23)$$

$$\frac{h_r d}{\lambda} = K \left(\frac{d^2 N_s \rho}{\mu}\right)^{0.67} \left(\frac{C_p \mu}{\lambda}\right)^{0.33} \left(\text{other factors}\right)^x \qquad (3\text{-}24)$$

where d is the vessel diameter, N_s is the agitator speed, and the physical properties relate to the reaction fluid. Other factor ratios depend on the vessel geometry and the viscosity characteristics of the reaction fluid at the reactor wall, and generally are not of significant influence. Note that the Nusselt number, which contains the vessel diameter as the dimension, is a version of the Biot number.

For scale-up, the relation of the plant system (plt) to the laboratory or pilot plant measurement system (exp), in a simplified manner, becomes:

$$h_{r\text{plt}} = h_{r\text{exp}} \left(\frac{d_{\text{plt}}}{d_{\text{exp}}}\right)^{0.67} \qquad (3\text{-}25)$$

This general method can be used for geometrically scaled reactors with Newtonian liquids or with Newtonian suspensions in the turbulent flow region ($N_{Re} > 1000$).

For non-Newtonian liquids and suspensions, an apparent viscosity is determined using correlations which include power input and the Reynolds number. Scale-up comparisons based on heat generation data only were determined by comparison of results from RC1 experiments and from a 675-liter reactor [208]. In the experiments, a Bingham plastic fluid was used to determine the film heat transfer coefficient. This presents a worst case because of the low thermal conductivity of the Bingham plastic. Calculated inside film heat transfer coefficients determined in the RC1 tests were about 60% lower than the values determined in the pilot plant reactor, even though substantial effort was made to obtain both geometric and kinematic similarity in the pilot reactor.

3.3.4.5 Self-Heating

Self-heating occurs if the heat generated by an exothermic reaction in a vessel cannot be removed from the system, regardless of whether the vessel is a reactor, a storage silo, or, for that matter, even a large pile of material. Self-heating can be caused by a decomposition reaction, by contamination, or by autocatalysis, but is often caused by a slow oxidation from reaction with oxygen in the air. In nearly all cases, the self-heating process can be described by the heat balance indicated in Equation (3-1). The temperature at which the materials are charged to the vessel, and the coolant (or ambient) temperature under which storage takes place affect both the occurrence of a runaway and the time-lag involved in the self-heating process. Since the heat balance in Equation (3-1) cannot be solved analytically, relatively simple models were

3.3 ACQUISITION AND USE OF PROCESS DESIGN DATA

introduced to estimate the critical temperature above which a thermal runaway will occur.

Two classical models have been described for runaway calculations in which the important difference between the two is in the degree of mixing. The first model, proposed by Semenov [165], applies to well stirred mixtures where the temperature is the same throughout the mixture. Heat removal occurs with a steep temperature gradient at the surface of the walls or coils, and is governed by the usual factors of area, temperature of coolant, and heat transfer coefficients. Case A in Figure 3.20 shows a temperature distribution by the Semenov model for self-heating.

The second model, proposed by Frank-Kamenetskii [162], applies to cases of solids and unstirred liquids. This model is often used for liquids in storage. Here, it is assumed that heat is lost by conduction through the material to the walls (at ambient temperature) where the heat loss is infinite compared to the rate of heat conduction through the material. The thermal conductivity of the material is an important factor for calculations using this model. Shape is also important in this model and different factors are used for slabs, spheres, and cylinders. Case B in Figure 3.20 indicates a typical temperature distribution by the Frank-Kamenetskii model, showing a temperature maximum in the center of the material.

The more recent Thomas model [209] comprises elements of both the Semenov and Frank-Kamenetskii models in that there is a nonuniform temperature distribution in the liquid and a steep temperature gradient at the wall. Case C in Figure 3.20 shows a temperature distribution curve from self-heating for the Thomas model. The appropriate model (Semenov, Frank-Kamenetskii, or Thomas) is determined by the ratio of the heat removal from the vessel and the thermal conductivity in the vessel. This ratio is determined by the Biot number (N_{Bi}) which has been described previously as hx/λ, in which h is the film heat transfer coefficient to the surroundings (air, cooling mantle, etc.), x is the distance such as the radius of the vessel, and λ is the effective thermal conductivity.

The Semenov model applies when the Biot number is close to zero, and the Frank-Kamenetskii model applies when the Biot number is large. The

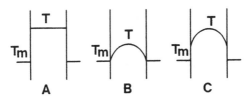

FIGURE 3.20. Typical Temperature Distributions during Self-Heating in a Vessel. A = The Semenov Model; B = The Frank-Kamenetskii Model; C = The Thomas Model

Thomas model is usually appropriate somewhere between these two. These criteria are guidelines only and must be carefully applied. All three models should be tested for borderline cases. In practice, the models are valid only if no mass flows to or from the vessels, negligible reactants are consumed, and heat is generated only by reactions.

The Semenov model, which applies to well stirred mixtures at a uniform temperature, is the basis of Equations (3-26) and (3-27) below:

$$\frac{1}{e} = \frac{Q\rho A \exp(-E_a/RT_m)}{(ShRT_m^2)/E_a} \tag{3-26}$$

$$1 = \frac{Q\rho A \exp(-E_a/RT_r)}{(ShRT_r^2)/E_a} \tag{3-27}$$

where S is the surface area equivalent to A_s in other expressions, and A is the pre-exponential factor in the Arrhenius equation. The two equations are different expressions for the same model in which the heat production rate is in the numerator and the heat removal rate is in the denominator. In Equation (3-26), T_m is the ambient or coolant temperature, and the critical value of the quotient is $1/e$ (about 0.368). In Equation (3-27), T_r is the temperature of the reaction mixture, and the critical value of the quotient is 1. The two temperatures are related by Equation (3-28):

$$(T_r - T_m)_{max} = RT_r2/E_a \tag{3-28}$$

The Frank-Kamenetskii model, which applies to solids and unstirred liquids, is represented by Equation (3-29) below. The heat production rate is in the numerator and the heat removal rate is in the denominator.

$$\delta = \frac{Q\rho A \exp(-E_a/RT_m)}{(\lambda RT_m^2)/(E_a x^2)} \tag{3-29}$$

Runaway will occur when the calculated delta (δ) exceeds the critical delta (δ_{cr}) which depends on the shape of the reaction mixture: 0.88 for a plane slab, 2.00 for an infinite cylinder, 2.78 for a right cylinder with l/d equal to 1, and 3.32 for a sphere. Bowes [133] provides formulas for calculation of δ_{cr} for other geometric shapes and structures. In this model, heat is lost by conduction through the material to the edge, where the heat loss rate is infinite relative to the conduction rate. In this model, there is a maximum temperature in the center as shown in Figure 3.20 Case B.

Examples of the use of the Semenov and Frank-Kamenetskii models are presented by Fisher and Goetz [210].

Thomas [209] describes the effect of Biot number on the critical δ in calculations of runaway temperatures. Biot numbers for right cylinders with various l/d ratios are available.

3.3 ACQUISITION AND USE OF PROCESS DESIGN DATA

These models describe simple vessels and shapes only. If the vessel structures become more complicated or if reactant transport determines the reaction rate, as is often the case with oxidation by air, more complex numerical models must be used [138].

3.3.4.6 Scale-up of Accelerating Rate Calorimeter (ARC) Results

The ARC is described in Chapter 2 (Section 2.3.2.3). Measurements can be made to determine dT/dt, dp/dt, and p_{max}. The important ΔH_d or ΔH_r can then be calculated. Using p_{max}, the gas production per unit mass can also be calculated. This value is used for estimating the pressure load for the plant unit in the vessel design and the plant layout design.

The time-to-runaway can be calculated using dT/dt and T_{max} values. This calculated time is a measure of the possible global reaction rate. ARC experimental results may also be used to develop required mathematical models for process design.

Kohlbrand et al. [131], present an example using ARC data for development of an intervention strategy for an uncooled storage tank. A reactive monomer mixture is continuously added to a 10-m^3 tank over a 6-hour period. The mixture remains nonviscous to about 40% conversion. The tank will be 40% full, resulting in an effective heat transfer area of 8 m^2. The measured ARC data are corrected for the measured phi-factor (from 1.63 to 1) and for an assumption of pseudo-first-order reaction. With adiabatic conditions, the temperature increase between 45°C (starting temperature) and 95°C takes about 675 minutes. The calculation is based on an initial self-heat rate (i.e., at 45°C, of 0.025°C/min).

Several strategies are then developed related to the temperature of the tank contents. The first strategy is to attempt to process the material in the tank in a normal fashion. A simple calculation shows that the maximum ambient temperatures (T_{mmax}) to maintain equilibrium at a given reaction temperature are:

at 90°C $T_{mmax} < 36°C$

at 95°C $T_{mmax} < 14°C$

It was concluded that this strategy will work at reaction temperatures below 90°C, provided the heat transfer coefficient is maintained.

Other strategies examined included the assumption of a total adiabatic condition and the effect of water addition on the slowdown of the reaction rate and hence the temperature rise.

3.3.4.7 Scale-up of Vent Size Package (VSP) Results

The Vent Size Package (VSP) apparatus was described in Section 3.3.2.6. The relatively small 100-cm^3 test cell size implies that the samples tested may not be fully representative of commercial scale systems in some cases. The contents of the test vessel may be stirred, although the stirring facilities are not very efficient. The data obtained from the VSP experiments can be used in combination with a variety of vent size design packages to obtain an estimate of vent dimension requirements for either one- or two-phase venting.

A simplified VSP procedure which enables a quick estimate of the required vent diameter is discussed in [191]. The venting characteristics in connection with runaway chemical reactions can be related to vapor and gassy (including hybrid) systems.

For a pure vapor system ($\Delta T/\Delta t$ driven), where the runaway reaction can be kept under control by the latent enthalpy of evaporation (tempered system), a relatively simple expression can be used for the estimation of the necessary vent diameter.

For some gassy systems ($\Delta p/\Delta t$ driven), viscosity considerations become important which make testing difficult if not prohibitive.

For gassy and for tempered systems, the flow rate can also be measured in a simulated vent line (same l/d ratio) of diameter d_0. Additional calculation formulas are given in [191].

For gassy systems, including hybrid systems where vapor stripping may be sufficient to control the runaway by latent enthalpy of evaporation, a pressure increase following a relief actuation is generally dominated by noncondensables. In other words, the partial pressure rise rate of the gassy material is much greater than that of the pure vapor, that is, $(dp/dt)_g >> (dp/dt)_v$. In this case, the overpressure is reached quickly causing the discharge to be dominated by nonflashing two-phase flow. For choked conditions, this results in an approximate expression for the vent diameter.

Another approach for scale-up of safety relief for runaway reactions is shown in Figure 3.21 and is discussed by Fauske in [211]. The following sequence is used:

- *Step 1*—acquisition of thermal data, e.g., the adiabatic temperature rise ΔT_a,
- *Step 2*—acquisition of mass flow rate data; some experimental precautions must be taken in order to obtain the proper data since the objective is to determine the two-phase critical flow rate by measuring the emptying time ΔT_E, and
- *Step 3*—size vent for large scale reactor.

A safe, but not overly conservative vent size for the large scale is given by the relationship shown in Step 3 of Figure 3.21, in which A_{LS} is the area of the scale vent line, A_T the area of the test vessel vent line, Δt_a the measured adiabatic rise time, ΔT_E the measured emptying time, V_{LS} the volume of the large-scale vessel, and V_T the volume of the test unit.

3.3 ACQUISITION AND USE OF PROCESS DESIGN DATA

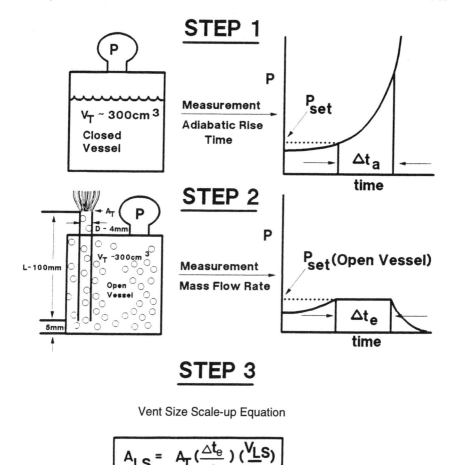

FIGURE 3.21. An Approach to Emergency Relief System Sizing in Case Necessary Kinetic and Thermo-physical Data Are Lacking.

3.3.5 Process Design Applications

Process design and design for safety go hand-in-hand. Therefore, safety testing is necessary and useful in the various stages of process development as illustrated in Chapter 1 (Table 1.1). The flow sheets covering test plans, strategies, and procedures discussed in Chapters 2 an 3 lead to a hazard evaluation of the substances and the reactions used in the process, an evaluation of the process design, and then of the final plant design.

The choice of reactor systems from the safety perspective are discussed by examples in the following sections.

3.3.5.1 Batch and Semi-Batch Processing Plants

Batch processes are characterized by [212]:

- non–steady state conditions,
- continuous variations in physical properties, chemical compositions, and physical state of the reaction mixture,
- sharing auxiliary equipment, such as columns and condensers, with other reactors, introducing potentially hazardous consequences if operating procedures are deficient,
- interaction between process and control system,
- checking that the process goes through all the correct processing steps, in the proper order, and at the proper time,
- the relatively large inventory in comparison with continuous operations,
- complex process piping which can introduce the risk of cross-contamination among different processes for multipurpose equipment,
- frequent start-ups which increase the probability for errors, and
- general overall design compared to dedicated design for a specific process.

Many of the above factors apply to semi-batch reactor systems as well. Also, the following factor is important for semi-batch reactor systems:

- changing material levels in the vessel as materials are charged and perhaps withdrawn, thus changing mixing characteristics and effective heat transfer area.

Good process and equipment design practices for batch and semi-batch operating systems, particularly for those involving reactive or otherwise hazardous substances, are:

1. *Minimize hazardous material inventories*
   - Substitute less hazardous materials where possible.
   - Use less hazardous forms of a required material.
   - Identify chemical interactions through an interaction matrix.
   - Generate hazardous unstable or toxic intermediates *in situ* from less hazardous starting materials, avoiding storage and transfer of the hazardous substances.

2. *Separate process steps*
   - Perform each step in a separate vessel with the design optimized for that operation; this includes vessel size, pressure rating, emergency relief system design, materials of construction, utilities, and agitation design.
   - Use fewer pipe connections to each vessel, reducing the risk of cross-contamination, inadvertent introduction of the wrong material, or the use of the wrong utility.

3.3 ACQUISITION AND USE OF PROCESS DESIGN DATA

3. *Design equipment to be versatile*
   - This includes variable speed agitation, segmented jackets or coils, oversized condensers, and oversized vapor lines.
   - Containment within the vessel for the credible worst-case scenario reducing the design requirements for the emergency relief system; this step is frequently too expensive and impractical in a multipurpose facility.

4. *Limit error potential*
   - For key raw materials that can create significant hazards if overcharged; the maximum possible charge should be physically limited by sizing the feed tanks to hold no more than the required quantity.
   - Feed raw materials via a metering pump in connection with a pump timer.
   - Limit steam pressure with a relief valve or pressure control valve on the steam connection, thus limiting the maximum temperature short of a runaway.
   - Use semi-batch processes for exothermic reactions and monitor the consumption rate of the limiting reactant.

5. *Follow prudent management control*
   - An effective management system to ensure process integrity is essential and must include up-to-date operating procedures, product changeover instructions, and checklists that cover instrumentation, equipment arrangement, and procedures; procedures must also include adequate checks and reviews to ensure the required changeover modifications have, in fact, been made when required.
   - A thorough understanding of the potential hazards of the substances involved and the process is essential, particularly taking into account the possibility of cross-contamination; am interaction matrix is a useful tool to obtain the needed information.
   - Identify and safety test critical equipment and procedures.
   - Equipment design should facilitate cleaning between batches of different products.
   - Thorough operator training is essential.

A number of hazard identification and analysis techniques (e.g., HAZOP), can be applied to identify, analyze, and reduce and/or mitigate the process hazards, which includes handling of reactive chemicals and energetic reactions. Chapter 4 provides an overview of these kinds of techniques as related to reactive chemicals; mote detailed reviews of hazards analysis techniques are included in [2, 3].

3.3.5.2 An Example Involving Peroxides
A technique is described [213] in which reactive chemical experimental data obtained with peroxides and hydroperoxides are used to define safe design

parameters. Peroxide processing hazards include thermal runaway, and liquid and vapor phase deflagration. Several small scale tests were conducted to define and quantify the hazards. Testing for the thermal runaway hazard was conducted in a VSP or similar equipment, and in the PHI-TEC. Three basic types of tests were run as noted below:

1. closed cell test to estimate the consequences of a nonvented runaway reaction, that is, to determine p, dp/dt, and dT/dt,
2. a rapid blowdown test to establish the vapor–liquid ratio likely to enter the relief line during a venting incident, and
3. a controlled test to determine the heat generation rate.

Typical results in the closed cell test for peroxides and hydroperoxides were a p_{max} of 110 bar and a T_{max} of 255°C. From the blowdown test, it was concluded that two-phase blowdown occurred. Furthermore, the peroxide concentration in the residue after blowdown was about twice that of the original concentration.

The tempered vent size testing results indicated that at 190°C, a tempering of the runaway occurred for some time during which the solvent evaporated. Then the temperature rose again, possibly due to the residual concentrated peroxide.

The liquid phase deflagration testing resulted in a temperature–concentration relation which divided the regimes at which the peroxide does and does not sustain a deflagration. This relation depended on the test tube diameter (2.5-cm and 7.6-cm tubes were used), and thus extrapolation of the diameter of interest is required.

The vapor phase deflagration was investigated in a 5-liter vessel at 120°C using a heated wire as the ignition source. A limiting pressure related to the peroxide or hydroperoxide concentration was obtained. Thus, for any given operating pressure, the maximum safe level of the peroxide or hydroperoxide to prevent deflagration in the vapor phase can be obtained, although a runaway still could occur at the temperature involved.

Design applications
It is possible to design vessels to contain vapor phase deflagrations since the maximum explosion pressure is less than a factor of 8 above the initial pressure. However, a vapor phase explosion may subsequently ignite a liquid phase explosion, which is what happens with 70–80% *t*-butyl hydroperoxide (TBHP) in water [214]. The heat evolved during the gas phase explosion heats up the interfacial area, resulting in evaporation of the peroxide moiety. This is enough to sustain a deflagration in the gas phase. In the TBHP/water case, the process is tempered by the presence of water. The presence of a high boiling solvent could lead to a deflagration in the liquid phase, producing considerably more gaseous products and thereby increasing the deflagration rate.

3.3 ACQUISITION AND USE OF PROCESS DESIGN DATA

Thus, another reason to stay out of the vapor phase explosion region is to prevent the ignition of a liquid phase explosion.

The usual mitigation approach to runaway reactions, that is, the application of a vent relief system, cannot be adapted in the hydroperoxide case since venting results in a more concentrated hydroperoxide for which a runaway cannot be stopped. Therefore, operating conditions have to be chosen in such a way that a runaway is prevented (inherently safe operation). Increased control at operating conditions can be achieved by optimization of the heat exchange or by segmented the reaction vessel. Basic measures to be taken are always directed at an improved control of pressure and temperature. After safe operating conditions have been defined and instituted, additional safety measures should be considered. Examples are on-line detection of the rate of temperature rise and the use of dumping or quenching techniques.

3.3.5.3 An Example Involving a Continuous Nitration

This example involves a continuous adiabatic nitration process for the manufacture of mono-nitrobenzene (MNB) [215] by the reaction of benzene with nitric acid in a CSTR system. The process is designed to be inherently safe. No external cooling is used, but the reaction mass is heated by the reaction itself to a temperature level controlled by the amount of sulfuric acid–water mixture circulating through the system. This acid actually acts as both a heat sink and as a nitration enhancer. If the sulfuric acid pumps fail, the nitric acid and benzene pumps are automatically shut off.

In the hazard evaluation of the process, it was found that exotherms occurred with MNB–H_2SO_4 mixtures at temperatures above 150°C. The initiation temperature and extent of the exotherm depend on the acid concentration. During normal operation, the temperatures in the continuous stirred tank reactors and in the continuously operated separator are between 135 and 148°C. However, operating simulation showed that for certain feed rates well out of the normal operating range, the temperature could reach 180°C and a runaway is thus possible.

The safety study was performed in several steps. Small scale tests with a DSC were run to determine the most hazardous situation. The exothermic onset temperature, the heat evolved from the reaction, and the time-to-maximum rate were determined. Then larger scale tests were performed for thermal stability, pressure rise rate, and vent sizing, which in part gave information about scale-up issues. Venting was simulated before actual testing. Defensive measures other than venting (e.g., rapid drainage of the reaction vessel and quenching were studied, as were the problems associated with atmospheric discharge.

DSC experiments performed in closed vessels, with a heating rate of 5°C/min on samples of 10 to 20 mg, showed that the enthalpy of reaction was from 410 to 1175 kJ/kg, a significant variation range. The corresponding adiabatic temperature rise related to the enthalpy of reaction results was about 200 to 580°C. This temperature range yields a pressure rise that would certainly be a problem to contain. The onset temperature of the exotherm under the test conditions was usually well above 200°C.

Worst-case analysis based on the DSC data, namely, the test with the lowest onset temperature, resulted in a graph showing the relationship between initial temperature and time-to-maximum rate under adiabatic conditions. For an initial temperature of 170°C, it would take 2 hours to reach the maximum rate. Venting simulation tests were undertaken on a larger scale to detect safe venting requirements for the separator and for the MNB hold tank. Several vent sizes were tested. It was found that a 10-cm rupture disc with a burst pressure 1 bar above the operating pressure was adequate.

A level control failure in the separator could result in a spill of acid into the MNB hold tank, and an exotherm could take place there. This situation was simulated by testing a two-phase mixture consisting of 90% organics and 10% spent acid. The test vessel loading was 50% (in practice, loading is below 25%), and the bursting pressure of the rupture disc was 6 bar above operating pressures to simulate a worst-case condition. It was observed that a 25 cm rupture disc with a burst pressure of 2 bar above operating pressure was adequate for venting the separator.

A thermal scan showed that the exotherm of the principal reaction can be significant if the system is neither controlled nor vented. From isothermal studies (i.e., experiments at constant temperature), time-to-maximum rate was determined which was comparable to that obtained from the DCS data. The larger scale data showed, not surprisingly, more rapid reactions at elevated temperatures. Thus, it was decided to use the DSC data at lower temperatures, and the larger scale test data at higher temperatures for hazard evaluation.

Vapor venting simulations were performed. Several parameters were varied in the simulation, such as up to a tenfold increase in the reaction rate and a doubling of the enthalpy of reaction. Failures of the control system and of the operators were simulated as well. It was concluded that the system can vent successfully, and that the rate of decomposition is not sufficiently rapid to allow for significant self-acceleration.

A mitigation action considered prior to venting is the rapid draining of the nitrators, separator, and crude MNB tank. The purpose of draining is to evacuate the vessels before significantly higher temperatures are reached in an emergency situation so that venting may be prevented. The liquids were discharged into a quench tank containing cold concentrated sulfuric acid. The time required to drain each unit was calculated and compared to the time needed to reach a runaway decomposition reaction. The calculations showed that each vessel could be drained in sufficient time. Actual drain times were checked during start-up, and these agreed well with the calculated values.

Quenching of the drained fluids was calculated as a function of the initial fluid temperature, and of the ratio between fluid and cold sulfuric acid. As a result of an automatic drain at 30°C above normal, the resulting temperature was found to be 90°C.

3.3 ACQUISITION AND USE OF PROCESS DESIGN DATA

Atmospheric dispersion of any rupture disc discharges would result in a vapor cloud with gas concentrations above the lower explosive limit. Thus, such releases must be avoided, and other mitigation procedures should be used. However, as an additional check on the situation, mapping of the potential gas cloud versus the plant layout was conducted with the conclusion that no ignition sources were likely to be present in the region where the vapor cloud would be flammable.

As a result of the experimental studies, the simulations, and the calculations, the following safety precautions were taken. The only foreseeable process upset resulting in a temperature excursion in the nitrators is a deviation in the feed ratios. Control features and interlocks were installed to reduce this possibility. The sulfuric acid flow control station was designed in such a way that flow of this process heat sink is not halted upon complete failure of the flow controller. Low sulfuric acid flow results in automatic shutdown of the nitric acid and benzene feeds.

The overall study resulted in using the a number of additional levels of protection against high temperature deviations in the plant.

3.3.5.4 A Self-Heating Example

This example involves the detonative explosion of 3, 5-dinitro-*o*-toluamide (dinitrolmide) which had been left inside a closed dryer vessel for a period of 27 hours after the drying process had been completed. The insulation on the dryer maintained the material at a temperature between 120 and 130°C. Under these conditions, the material began to decompose with the evolution of heat, causing self-accelerating decomposition, which led to a detonation. From the results of hazards testing performed on the material prior to production, conclusions had been reached that the material could be safely processed at the planned operating temperatures. The DSC measurements using freshly prepared material showed exotherms starting at a temperature of 273°C, and, using somewhat aged material, at 248°C. These results had not been alarming because these temperatures were well above the normal processing temperatures.

Other screening tests, including shock sensitivity and flammability tests, and thermodynamic computations raised no specific concerns. After the explosion, the material was tested in an ARC. Such testing showed that a typical batch of the compound could self-heat to full decomposition if held under adiabatic conditions at 120 to 125°C for 24 hours. These tests were run within the normal processing temperature range, and the ARC test results were hailed as demonstrating the likely cause of the accident.

It is important to note that the error of method in this case was not in using the DSC for hazard testing instead of the ARC, but in not checking for autocatalytic reactions in the initial testing. Any exothermic reaction will exhibit self-heating in various tests and will certainly run away under adiabatic conditions, but only a few reaction types (for example, autocatalytic and

154 3. CHEMICAL REACTIVITY AND PROCESS/REACTOR DESIGN

inhibitor depletion) will show increasing reaction rates at constant temperature. The error in testing was in not checking for autocatalysis (for which nitro compounds are notorious) by running the DSC isothermally. The increase of heat evolution with time would have called attention to the autocatalytic reaction. Not all self-heating reactions are caused by autocatalytic or self-accelerating mechanisms. It is true that the DSC in the scan mode will not reveal an autocatalytic reaction, but in the isothermal mode, it will show an increasing heat evolution that identifies autocatalysis. In the case described, the fact that the "somewhat aged" material showed an exotherm at a lower temperature than the freshly prepared material should have raised questions about autocatalysis, suggesting further tests.

3.3.5.5 Batch-to-Continuous Example
Originally, nitroglycerine was manufactured by batch process. This represented a significant hazard because literally tons of product and spent acid were maintained for several hours at elevated temperatures. In an attempt to reduce the hazard, the operation was changed from batch to continuous, a process in which the glycerine and nitrating mixture were separately fed into a reaction chamber. In this way, the residence time was reduced to several seconds, which obviously resulted in a safer operation.

3.3.5.6 Integrated Relief Evaluation
SAFIRE (Systems Analysis for Integrated Relief Evaluation), a computer program, and the accompanying users' manual [216] are important design tools which can be used to estimate the pressure–temperature–time history for a vessel containing a runaway reaction, or exposed to fire, based on experimental data and thermodynamic calculations. The program can be used to determine the necessary vent size to keep the pressure below a specified value. The system uses a reaction modeling approach. It can handle such aspects as:

- venting of nonideal gases, two-phase vapor-liquid mixtures, or non-condensible liquids,
- transition from two-phase flow to vapor-only flow (partial disengagement),
- complex reaction chemistry, and
- nonideal vapor–liquid equilibria.

3.3.6 Storage and Handling

3.3.6.1 Scale-up Example for Storage
Liquids and solids are frequently handled in bulk at ambient temperatures. The heat generated by thermally unstable materials is generally quite low under such conditions. However, because of the large masses involved, even

3.3 ACQUISITION AND USE OF PROCESS DESIGN DATA

these low heat generation rates may lead to self-heating and self-ignition of the bulk because of the very low rates of heat loss.

Small-scale simulation may not be possible because rates of heat loss corresponding to the large scale cannot be realized experimentally. In practice, two types of approaches are used to obtain reliable data for scale-up: (1) direct measurement of heat generation at storage temperature in extremely sensitive calorimeters (microcalorimetry), and (2) extrapolation of data that are obtained in investigations at temperatures higher than the conditions of bulk storage. A practical limit exists in the first approach because of the relation between heat generation and the size of the two systems. In the latter approach, oxidation or decomposition reactions are likely to change with temperature and, thus, can interfere with the validity of the extrapolation, usually using the Arrhenius method. Moreover, practical oxidation situations and typical ventilation systems are difficult to test in this latter approach.

A comparison has been made between small scale test results and a field trial at a 17-ton scale for a solid compound [217]. The test results from a very sensitive calorimeter (Thermal Activity Monitor from ThermoMetric, Sweden) were substituted in a model. and the self-heating situation in bulk containers was predicted. The large-scale trial was carried out in a steel rectangular container lined with polyethylene. A control device was used to keep the container at a temperature of 40 to 45°C. Several thermocouples enabled monitoring of the temperature as a function of time in different places in the large container.

The following conclusions were drawn. Microcalorimetry can be used to evaluate the heat generation characteristics of a solid material directly at the temperature of practical interest. However, in order to determine the worst case, the variability between batches of the substance must be determined which requires a considerable number of tests (over 100 trials in this specific case). Having obtained the heat generation as a function of temperature for with worst case, the safe storage diameter of storage vessels can be calculated and the equipment appropriately designed.

3.3.6.2 Peroxides

Another example is the safe handling of peroxides on a commercial scale. Earlier in Chapter 3 (Section 3.3.5.2), the design of a process plant handling organic peroxides was discussed [213]. Special emphasis is presented in on storage and handling of peroxides during manufacture. As discussed previously, organic peroxides have a number of particular properties in common, such as:
- sensitivity to heat,
- sensitivity to contamination,
- generation of heat on decomposition,
- formation of gases and mists on decomposition (selective),
- formation of free radicals on decomposition, which further catalyze decomposition, and
- limited oxidizing properties.

The results of a number of tests such as those described in Chapter 2 led to classifications for the peroxide group. These tests included the determination of the hazards of decomposition (deflagration and detonation), burn rate, fire hazard, and reactivity hazards. Five different classes were formulated, as listed in the NFPA 43B Hazard Class, from the test results. Emergency procedures have been established for these five classes.

The most important parameters to be controlled, whether in shipping or storage, are temperature, contamination, confinement, and quantity. During shipping, storage, and handling, the temperature of the peroxide should be kept well below its self-accelerating decomposition temperature (SADT) by temperature control techniques. By preventing runaway decompositions, the evolution of gases and mists (risk of vapor cloud explosion), autoignition, and loss of product can be avoided. Decomposition kinetics are also required for design of safe storage and handling equipment and procedures.

Contamination of peroxides has been a major source of accidents by runaway decompositions, particularly during handling and use. Therefore all equipment that is in contact with peroxides must be thoroughly cleaned. When a diluent is used, it should be properly selected and its purity must be strictly controlled. The use of dedicated loading/unloading equipment and avoidance of the use of shared or manifold equipment are methods to reduce the possibility of cross-contamination.

Once decomposition is initiated, confinement of an organic peroxide (or a formulation containing a peroxide) can strongly accelerate the rate of decomposition. Transition can occur from a controllable decomposition to a deflagration, which can no longer be vented, if the peroxide or peroxide formulation has a high decomposition energy. Dilution of the peroxide may, therefore, be required. Quantities in one place should be kept to a minimum. Sufficient venting should be provided to prevent a pressure increase which would accelerate decomposition. The quantity in storage must be controlled in accordance with the limitations in NFPA 43B. During use, the quantity should be maintained at the lowest practical level to minimize any hazards from decomposition or fire.

Measures must be taken to prevent spills. Further, it is necessary to know what action is recommended in case of spillage for each specific formulation. The disposal of waste needs special attention as well.

Adequate fire protection must be provided by separation, walls, and/or sprinkler control (NFPA 43B) in order to prevent damage to the surrounding areas in case of a decomposition or fire.

3.3.6.3 Passive Means to Prevent Explosions

The term *explosion* in a storage and handling sense usually implies the rupture of a vessel. Fauske [191] reviewed the hazards in the chemical industry in connection with storage and processing of chemicals. The major hazards discussed are a boiling liquid expanding vapor explosion (BLEVE) for high-

3.3 ACQUISITION AND USE OF PROCESS DESIGN DATA

pressure storage of flammable gases, runaway of a reactive material at low-pressure storage, and two-phase flow for unconfined vented runaways during chemical processing. Another major hazard is the unconfined vapor cloud explosion [218]. Reference can be made to the AIChE/CCPS *Guidelines for Evaluating the Characteristics of Vapor Cloud Explosions, Flash Fires, and BLEVEs* [159].

The conditions that promote a BLEVE are external fires heating up the tank walls above the liquid surface. This heating weakens the tank wall surrounding the vapor phase leading to a rupture. The released liquid flashes and ignites, resulting in a huge fireball.

Following are some examples of passive safety systems to reduce the likelihood of explosions in storage units. The use of baffles in a high-pressure storage vessel can cool the tank wall above the liquid surface via liquid pumped around by vapor bubbles, extending the time for fire fighting. Fire resistant tank insulation is also effective in delaying a BLEVE.

A solution to safe high-pressure storage is the double-walled tank [191]. The space between the two walls is filled with water, provided the stored substance is not sensitive to water in case of a leakage. A relief valve pipeline ends beneath the water surface so that either absorption or entrainment of the vapor takes place. The water-filled section of the tank has an open vent to the atmosphere.

A similar type of vessel can be used for the atmospheric storage of flammable gases as well. A typical design is shown in Figure 3.22. The water level in the secondary vessel should be such that the top of the storage vessel is always covered. In this way, the hazard from fire exposure is prevented and the collapse of the inner vessel is largely eliminated by minimizing any hydrostatic forces. A small vent could lead directly to the atmosphere for nontoxic substance storage. For toxic materials, the vent discharge could run beneath the water surface if the toxic material is soluble in and not reactive with water.

These proposed passive safety systems minimize the need for active safety installations in storage, such as the provision of water cooling or water curtains.

3.3.7 Dryers and Filters

Many fires have occurred in industrial dryers and filters [138, 218]. Typical causes include ignition by electrostatics and self-heating of deposited layers. The combination of elevated temperatures, air, combustible material, and thick layers (which insulate the heated surfaces) have led to runaways, ignition at hot spots, and fires. Figure 3.23 illustrates a case of self-heating of a layer of solid material. The runaway may be caused by decomposition and/or oxidation reactions. The damage by self-heating will result in loss of product or reduced product quality if the heat generated is ultimately balanced with the

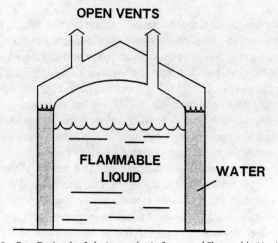

FIGURE 3.22. One Design for Safe Atmospheric Storage of Flammable Liquids.

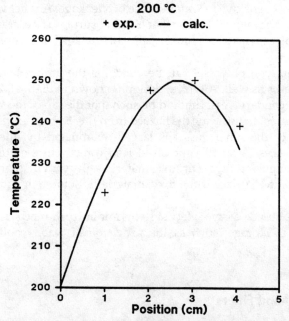

FIGURE 3.23. Calculated and Measured Temperatures in a Layer as a Result of the Self-Heating of Tapioca.

heat exchanged. Since this is often not the case, the product will self-heat to the point of decomposition or self-ignition, which may result in an explosion.

A layer of particles may be deposited on a peripheral surface, for example, a steam pipe or the outer wall of a vessel. The particles are then exposed on the surface area and by the gas phase to temperatures that are generally high.

3.4 PROTECTIVE MEASURES

The period of time during which the particles are exposed well exceeds the standard residence time in the equipment since such layers are not generally removed during the production run. Self-heating can occur.

Both self-heating and self-ignition characteristics are commonly determined by adiabatic tests (self-heating) and hot-plate tests (self-ignition). These tests are limited in their applicability because they do not fully simulate the plant conditions.

A mathematical model is described [138] in which the self-heating of material layers under industrial conditions is simulated. The model takes into account oxygen (or gas) diffusion and consumption, reactant conversion, heat conduction in, and heat transfer to and from the layer. Scale-up experiments were performed which showed the model can be successfully applied to predict the self-heating phenomenon in the layers.

A general approach to evaluating decomposition and oxidation of product during drying follows:

1. Determine the lowest temperature at which exothermic decomposition results in heat accumulation.
2. Investigate the oxidation reaction as a function of temperature and oxygen concentration (for air drying only).
3. Determine a residence time for the product at the chosen conditions taking into account the results of investigations mentioned previously.
4. Use the model to determine the time at which deposited material may show a runaway, defining the time at which the dryer must be cleaned.
5. Consider friction and hot spots in making the evaluation.

A typical approach for safety precautions to be taken to prevent fires and explosions in handling solid materials is presented by Reay [219].

3.4 PROTECTIVE MEASURES

This section on protective measures discusses three elements: (1) containment, (2) instrumentation and detection of a runaway, and (3) mitigation measures. For each element, examples are given to illustrate the principles discussed. This section is basically a summary of protective measures, not an exhaustive treatise. Protective measures are necessary considerations, and in fact, safety requirements, when handling reactive substances and exothermic reactions.

3.4.1 Containment

3.4.1.1 Introduction
The approaches for containment of runaway situations will differ depending on several factors, for example:

- laboratory, pilot plant, or plant situation,
- normal heat and gas releases versus inadvertent releases,
- containment area versus containment vessel,
- required extent of containment (or conversely, acceptable releases) based on severity of explosibility or toxicity, location, and other factors,
- total containment versus nominal containment, that is, containment having limited releases with or without effluent release collecting systems.

High-pressure reactions, including high-pressure polymerizations, may and sometimes must be run under total containment. A high-pressure polymerization will probably not lead to a deflagration in case of a runaway upset. Working with explosive substances or high-pressure test procedures may be subject to containment regulations. If the substances involved or the effluents likely to be released are highly toxic, total containment must be considered. In the latter case, it may be possible to collect the relief system effluent in another vessel, where the contents can be held safely for subsequent treatment.

In the approach to acceptable containment design, some other important questions are:

- How much gas will be released?
- How toxic is the gas?
- What is the likely physical state of the vented material (one or two phases)?
- At what temperature is the release likely to be?
- What effect will occur—thermal explosion potentially followed by a deflagration, or is a detonation also possible?
- How large is the free volume in the vessel (the smaller this volume, the larger is the likely effect)?

Total containment is not a practical alternative if there is a potential for detonation or even a rapid deflagration.

A vapor–air or dust–air deflagration may be contained if the design considers the internal pressure in the vessel caused by the deflagration followed by a controlled release of the gases of combustion, and the possible external pressure that could result in an implosion after venting of the deflagration gases. The same requirements are valid for auxiliary equipment. In many cases, design for containment is generally applicable to pressure build-up from runaway reactions, which thus results in an inherently safe design. This concept was investigated using the DIERS technology [221, 222].

3.4.1.2 Determination of Gas–Vapor Release

Several test methods are available to determine the amount and temperature of the gases that will be released during a runaway, for example, the the ARC, the closed bomb, or the high-pressure autoclave. In these test methods, a small

3.4 PROTECTIVE MEASURES

amount of the substance or reaction mixture being investigated is weighed to give a ratio of substance mass to autoclave or bomb volume between 0.1 and 0.25. The material in the test vessel is heated to the desired temperature, and the pressure is then measured as a function of time. A relation exists, as shown in Equation (3-30), between the mass of the substance and the maximum pressure which is determined for both thermal explosions and deflagrations:

$$F/p = m/V_b + C \qquad (3\text{-}30)$$

where F is a factor that is a constant for the given substance. It accounts for both the amount and number of gaseous products, and the temperature of the gases at maximum pressure.

Concentrated organic peroxides, such as t-butyl peroxybenzoate (TBPB), have an F value of about 100 to 150 kJ/kg. These compounds can produce a runaway ending in a deflagration. Dilution of the peroxide with the proper solvent will result in a considerable decrease of the F value because of the decrease in concentration of the active component and the decrease in the maximum temperature due to heating and evaporation of the solvent.

3.4.1.3 Laboratory Scale

This section reviews the criteria for hazards testing of reactions on a small scale, particularly whether the experiments should be run in an open laboratory or in a high-pressure cell.

Serious consideration should be given to hazard testing a reaction under containment conditions for reactions that are known to be energetic, that are conducted at elevated pressures, or that involve well-known or anticipated significant hazards. Examples are hydrogenations, nitrations, polymerizations, halogenations, oxidations, and rearrangement reactions. Reactions that involve new chemistry, high temperatures, or toxic or corrosive materials are also candidates for testing under containment facilities.

The first step is the evaluation of thermodynamic and kinetic data by quantitative energy calculations and qualitative considerations as discussed in Chapter 2. The results may provide a satisfactory answer as to whether the reaction can be performed in the open laboratory or requires a high-pressure cell arrangement on the small scale. Further evaluations are required for scale-up. Toxicity, corrosivity, type of apparatus, size, and other criteria must also be considered.

Thermodynamic calculations can be performed using CHETAH. This program predicts the maximum reaction energy of chemical compounds and is useful for the preliminary screening of potentially hazardous substances or mixtures. CHETAH provides estimates of enthalpy, entropy, and heat capacity for many organic and organometallic compounds between 300 and 1500°K. It also computes the net change in enthalpy, entropy, and free energy for balanced chemical equations. Information about CHETAH was provided previously in Section 2.2.3.3.

3. CHEMICAL REACTIVITY AND PROCESS/REACTOR DESIGN

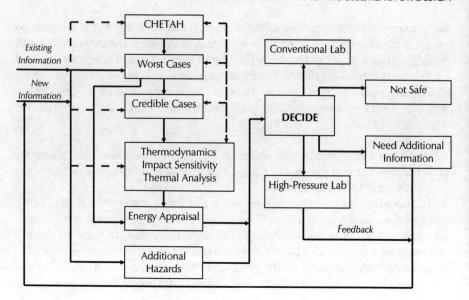

FIGURE 3.24. Flow Sheet to Determine Proper Site for Reactivity Testing (Laboratory or High-pressure Cell).

At this stage, potential worst case and credible case accidents must be evaluated as illustrated in Figure 3.24. The worst case may be the explosion of the reactor, releasing high-pressure gases and metal fragments. The explosion could be caused by a simple runaway reaction, but the maximum energy would be released when the reactants undergo total decomposition. A vapor cloud explosion is possible as well if organic vapors in the presence of oxygen ignite. The ignition may be caused by a decomposition or runaway reaction leading to very high temperatures. In the latter case, the energy release may surpass that of a decomposition.

Credible cases are identified when the probability of decomposition is low. Energy calculations of known or proposed chemical reactions and side reactions are carried out to determine a more likely level of energy release than the worst-case scenario. Therefore, it is necessary to define the most energetic reactions. Enthalpies of reaction are calculated, followed by calculations of the adiabatic temperature rise of the system and the corresponding pressure rise.

Following through the chart in Figure 3.24, the thermodynamics, impact sensitivity, and thermal analysis are defined. If the energy release potential calculated via CHETAH is higher than 700 kcal/kg, the reaction system is considered a definite hazard.

The sensitivity to impact can be determined as described in Chapter 2. Impact sensitivities below 60 J for solids and 10 J for liquids are considered positive hazards.

3.4 PROTECTIVE MEASURES

Thermal analysis can be carried out also as described in Chapter 2. An exotherm is considered a real hazard whenever it occurs within the extremes of temperatures expected in running the reaction.

There are eight different combinations of outcomes from the thermodynamics, impact sensitivity, and thermal analysis considerations. Each combination has unique characteristics, and hazard progression can be established (see Table 3.6).

Numbers 7 and 8 in Table 3.6 represent high hazard substances such as azides, peroxides, perchlorates, and nitro compounds. The handling of such materials requires extreme care and safety precautions.

Number 1 represents the least hazardous group.

On the basis of the test and calculation results illustrated in Table 3.6, it is possible to make an energy appraisal for the specific reaction under investigation. Generally speaking, the potential for a worst-case decomposition becomes greater when the hazard class is high as listed in Table 3.6. The worst case should be assumed instead of the credible case, in general, as follows:

1. the material is sensitive to impact (Numbers 3, 6, 7 and 8) or falls into Numbers 5, 6, 7, or 8, and
2. any credible case approaches a decomposition temperature determined buy DSC or ARC (within 100°C).

For flammable systems, the risk of explosion must be considered if the system is not operated under an inert or reduced oxygen environment. It should be noted that the possibility of equipment failure and the consequences of the failure increase with increasing operating pressures.

A laboratory pressure safety program [223] presents a list of questions helpful in evaluating the problem of hazard studies in the open laboratory

TABLE 3.6
Combinations of Parameter Sensitivities

Number	Thermodynamics	Test Results: Impact Sensitivity	Thermal Analysis
1	−	−	−
2	−	−	+
3	−	+	−
4	+	−	−
5	+	−	+
6	+	+	−
7	−	+	+
8	+	+	+

versus containment. The reference includes checklists to arrive at a safe design of laboratory pressure systems as well as a list of design requirements.

3.4.1.3 Full Scale Example

A process is described [224] in which an exothermic reaction takes place in a semi-batch reactor at elevated temperatures and under pressure. The solid and liquid raw materials are both toxic and flammable. Spontaneous ignition is possible when the reaction mass is exposed to air. Therefore, the system must be totally enclosed and confined in order to contain safely any emissions arising from the loss of reactor control, and to prevent secondary combustion reactions upon discharge of the materials to the atmosphere. Further, procedures and equipment are necessary for the safe collection and disposal of solid, liquid, and gaseous emission products.

The process is run in a semi-batch mode, and multiple reactors are used. There are several possible causes for a loss of control such as insufficient heat removal and loss of agitation. Overpressurization leading to the bursting of rupture discs takes place several times per year, indicating both the clear need for containment but also a need to consider design and control improvements. The reference describes the autoclave rupture disc assembly, procedures for replacement of the discs, the cleaning of the containment vessels, and the routine maintenance procedures for the containment vessels.

3.4.2 Instrumentation and Detection of Runaways

3.4.2.1 Methods of On-Line Detection

Several techniques are now available for on-line detection of either the start of a runaway reaction or a point very early in its development. These techniques include temperature changes specifically programmed and/or the first and second derivative of the temperature. The latter are especially valuable for multipurpose plants in which the operating temperature is frequently changed. Temperature alarms and power detection can be used to alert operators for conditions that can lead to a runaway. Fail-safe systems are used to mitigate the effects of a runaway. These systems and procedures include rapid discharge, dilution with inert material, high-pressure water deluge, and agitator control.

The On-Line Warning System (OLIWA) has been described [225]; it was introduced into the United States in 1986 [226]. Another detailed description is in [227].

Off-line measurements differ significantly from on-line techniques. Off-line measurements include safety testing in a laboratory, which has the major advantage of providing available time to redesign the process and/or equipment if a risk is identified. Of course, immediate process response is not possible.

3.4 PROTECTIVE MEASURES

The most important on-line methods are discussed below along with the hazard criteria used by each technique.

1. Measuring a single variable and checking against a preselected value—
The temperature hazard criterion is defined as the point at which the system temperature exceeds a specified value, or:

$T_{reaction} > T_{lim}$ (preselected value).

The most common supervision parameter is temperature, but pressure is a possible choice as well. Several other variables, such as level, pH, or physical property changes, can also be chosen since they are easily measurable, but these characteristics are usually important for purposes other than identification of thermal hazards. The temperature criterion method depends strongly on the knowledge of the process and is, therefore, generally not suitable for detection of unexpected dangers.

2. Supervision of the rate of temperature change—
Here the hazard identification criterion is the point at which the rate of temperature rise exceeds a specified value, or:

$(dT/dt)_{process} > (dT/dt)_{lim}$ (preselected value).

For this method, the first derivative of the temperature has to be determined from process measurements with amplified noise filtered out. Since the "safe" temperature need not be specified, the independence and selectivity of this method is greater than with the temperature criterion alone. Another advantage is that a potentially unsafe condition can be identified in its early development stage. However, a number of frequently used, but low hazard thermal processes are characterized by fairly high heating rates, making the use of the first derivative ineffective.

3. Supervision of the acceleration of temperature increase—
Another hazard identification criterion is the use of the second derivative of the temperature, that is:

$(d^2T/dt^2)_{process} > (d^2T/dt^2)_{lim}$ (preselected value).

The predictive ability with this method is higher than the previous two methods cited, but a high value of the second derivative can also occur during relatively low hazard processing.

4. Detection of the progressive increase of heat evolution—
The hazard identification criterion defined here is a second derivative involving the rate of heat evolution as:

$(d^2Q/dt^2) > 0.$

This method is implied in the OLIWA system, and is based on the following simple heat balance:

$$m(dQ/dt) = mC_p(dT/dt) + UA_s (T - T_m) \qquad (3\text{-}31)$$

For the purpose of hazard recognition, it is sufficient to check the following expressions:

$$(d^2T/dt^2) > 0 \quad \text{and} \quad d(T - T_m)/dt > 0.$$

Thus, for hazard identification, only the measurement of one or two temperatures is necessary. Actually, for equipment without a heating or cooling system, evaluation of the term (d^2T/dt^2) greater than zero is sufficient. The method is independent of detailed process knowledge and, generally, of human judgment.

5. *Monitoring of heat balance in batch processing—*
A technique is described [228] for solving a set of dynamic material/energy balances every few seconds in real time through the use of a minicomputer. This dynamic thermal analysis technique is particular useful in batch and semi-batch operations. The extent of the chemical reaction is monitored along with the measurement of heat transfer data versus time, which can be particularly useful in reactions such as polymerizations, where there is a significant change in viscosity of the reaction mixture with time.

6. *Recipe-based supervision—*
In general, hazard identification criterion represents the deviation of one or more measured variables from specified values. This is the basis upon which a significant percentage of risk analyses are done. For a chemical process, a number of measurable variables, physical properties, and states or positions of various parts of the overall equipment, e.g., pumps, valves, and motors, can be specified for every time or phase of the process. Certain deviations from the "standard" recipe or settings can then be defined in advance as hazardous, and thus can be used for initiation of an alarm at the early stage of a runaway or upset condition.

7. *Model based on-line identification—*
Various hazard identification criteria can be defined if sufficient knowledge of the process and of the equipment is available such that a mathematical model of the overall process can be constructed. The model can then be used to detect a hazardous situation, such as a runaway, developing at an early stage. This technique has potentially the highest predictive power, but does require an extensive knowledge of the chemical process and of the equipment characteristics. However, fully sufficient models are rarely available and their development is time consuming.

3.4 PROTECTIVE MEASURES

3.4.2.2 Methods of Noise Suppression

Spence and Noronha [227] use the same set of equations discussed in Section 3.4.2.1 for the detection of progressive increase in heat evolution, where the following conditions must be met: (1) the rate of temperature rise must be increasing, and (2) the rate of change if the temperature difference between the cooling medium and the contents must be positive, that is,

$$(d^2T/dt^2) > 0 \quad \text{and} \quad d(T - T_m)/dt > 0.$$

If both conditions are met in a cooled vessel, or if the first condition occurs in a noncooled vessel, the potential situations for a thermal runaway are present.

With reference to Figure 3.25, the detection of an upset or runaway should take place in the unstable region of the reaction to allow sufficient time so that appropriate measures can be taken to prevent or mitigate the actual runaway. Such measures include decreasing or shutting down the reactant feeds in a semi-batch or continuous reactor, use of emergency cooling, injection of an inhibitor, or quenching by adding a suitable solvent such as water. Mitigation measures are discussed later in Section 3.4.3.

The determination of derivatives incorporates the use of noise suppression systems. Three methods are discussed in [227]. The method of direct estimation by adaptive filters is stated to give a high degree of smoothing under normal operating conditions to avoid false alarms, and to provide good responsiveness when abnormal conditions occur, thus allowing an early warning. The method also provides standard error estimates that can be used in formulating decision rules. The use of standard error eliminates the occurrence of false alarms because the use of an arbitrary margin, such as 0.01°C/sec on temperature rise, still can present an uncertain level of risk.

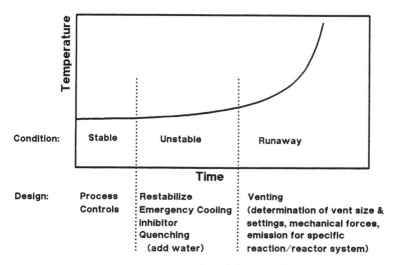

FIGURE 3.25. Concept of Restabilization and Venting

3.4.3 Mitigation Measures

3.4.3.1 Reaction Quenching Methods

In an early stage of warning for a runaway reaction, emergency cooling should be applied in an attempt to control the situation. This often is not sufficient to be a fully preventive measure, particularly if the runaway initiation was not identified quickly. Quenching of the reaction is the usual next step to take. The reaction can be quenched by using one or more of the following procedures:

- the rapid addition of an inhibitor for the given reaction system,
- the rapid addition of a compatible cold liquid, and/or
- the dumping of the reactor contents into a tank filled with a compatible cold liquid.

The last measure to be taken, after other methods fail, is venting. However, venting may cause pollution of the environment and/or potential injury to operating personnel or the public.

Some details of the quenching methods follow.

1. Addition of an inhibitor—

Inhibitors can be injected into the system in order to kill active species present, for example, by neutralizing the catalyst or by capturing free radicals in a polymerization. For example, the Lewis acid, BF_3-complex can be killed using gaseous NH_3 since the inactive compound $BF_3 \cdot NH_3$ is formed, and the reaction stops for lack of active centers. An antioxidant such as hydroquinone can be used to capture peroxide radicals to control reactions involving vinyl-type monomeric substances.

2. Addition of cold liquid (inert material)—

The temperature of the reaction mass can be decreased through direct heat exchange by the addition of a relatively cold liquid. The overall decrease in temperature leads to a lower reaction rate, and the active quantity of the reaction mass is decreased, which reduces the energy production per unit mass. This implies that some kind of mixing is available. Of less importance is the solubility between the cold liquid and the reaction mass. Thus, if water is added to an organic solvent which has a specific gravity less than 1, and the water does not dissolve in the reaction mass, the ultimate effect may be less than desired if mixing is not adequate.

Assuming a completely adiabatic system, the amount of cold liquid to be added (m_m) depends on the specific heat (C_{pm}) of the liquid, and the desired final temperature (T_E) according to:

$$mC_p(T - T_E) = m_m C_{pm}(T_E - T_m) \qquad (3\text{-}32)$$

3.4 PROTECTIVE MEASURES

Once the cooling liquid has been chosen (i.e., C_{pm} fixed), the variables of mass and temperature of the cold liquid remain and depend on the desired temperature decrease $(T - T_E)$. This value can be derived from the thermokinetics of the given reaction system.

The reaction volume must be large enough to contain both the original reaction mass and that of the added cold liquid. Furthermore, the boiling points must be considered since excessive boiling may overwhelm the condensing or venting system.

3. Dumping (rapid reactor discharge)—
Rapid discharge can take place through a bottom outlet by gravity or by pressuraztion of the vessel. Important factors are:

- the dumping rate in relation to the rate of temperature rise,
- the cooling of the dumped vessel contents by either on-line cooling or injection into cold liquid, and
- possible changes in flow characteristics upon cooling.

From the temperature in the system and thermokinetic data, the time available to the maximum rate of the runaway can be calculated or determined experimentally. This represents the maximum time available to discharge the contents of the reactor. The actual time needed for dumping should be less than the maximum to add a safety factor. The driving force for dumping can be gravity or vessel pressure.

The discharge flow rate of a liquid through a pipe is a function of the square of the discharge pipe diameter and the square root of the pressure difference between the reactor and the dump vessel. The dump vessel should be provided with a line to vent displaced gases (e.g., air, nitrogen). Care must be taken that the emergency lines can withstand the forces related to the initial liquid load. Toxicity of gases and vapors in, or evolved from, the dumped reaction mass must be considered in the design.

Dumping may be more practical for a small continuous reaction system than for a large batch or semi-batch reaction.

3.4.3.2 An Example Involving a Sulfonation

An example for the design of fail-safe systems for the continuous sulfonation of an aromatic compound has been described [229]. This investigation was undertaken because a thermal explosion had occurred in a pump and circulation line. The total exothermic decomposition energy of the reaction mass is 500 kcal/kg, which is large.

The potentially dangerous situations in the process were identified to be:

- *overheating*—too high a wall temperature, decomposition possible,
- *underheating*—too low a wall temperature, leading to crystallization and potential loss of heat transfer,

- *heat accumulation*—insufficient heat transfer, and
- *contamination*—water leakage could pose a potential hazard because of the high enthalpy of dilution.

The following fail-safe options were considered for the several vessels:

- rapid discharge (dumping),
- dilution with inert material (inert material in an overhead tank to be discharged by gravity into the vessel), and
- high-pressure water deluge (water is force-sprayed under pressure into the vessel, with vessel overflow into a catch basin).

A thermal stability study was first carried out to determine the following information: (1) the solidification temperature as a function of the concentration of the sulfonate; (2) the enthalpy of decomposition by DTA; (3) the autocatalytic nature of the decomposition by Dewar flask; (4) kinetic data for decomposition by Dewar flask; (5) the time to maximum rate by ARC, and (6) the heat generation as a function of temperature, also by ARC. In addition, the enthalpy of dilution was determined for various potential water leak rates. These data were useful in defining emergency response times.

Rapid cooling by flooding the reaction mass by an inert material was determined to be the best mitigation measure. This case included the dumping of 100% H_2SO_4 as the diluent into 4 m^3 of sulfonation reaction mass if a temperature of 30°C above the processing temperature (i.e., $T^p + 30$) was reached. A 4-m^3 tank was placed an appropriate distance above the vessel to be diluted. The heat generation was assumed to be caused by decomposition, and the heat absorption was determined through the heat capacity of the combined reaction mass and the 100% H_2SO_4 diluent. Before dilution, the temperature was allowed to rise to 30°C above the normal reaction. It was calculated that the sulfonation mass could be cooled back sufficiently rapidly to T_p by adding 1.54 m^3 of acid in 3 minutes through a 5-cm line. Although H_2SO_4 is denser than the reaction mass, cooling can still be accomplished under loss of agitation if dispersion is relatively uniform.

It is important to note that if 95% H_2SO_4 is used, a larger mass of diluent is required because of the presence of water.

3.4.3.3 Relief Disposal

The last defensive measure to control a runaway is to vent the gaseous and liquid products present in the system in a safe manner. The vent line may end in:

- a dispersion system,
- a flare system,
- a scrubbing system,
- a collecting system (catch tank), or
- the environment (which should be done only in an extreme emergency).

3.4 PROTECTIVE MEASURES

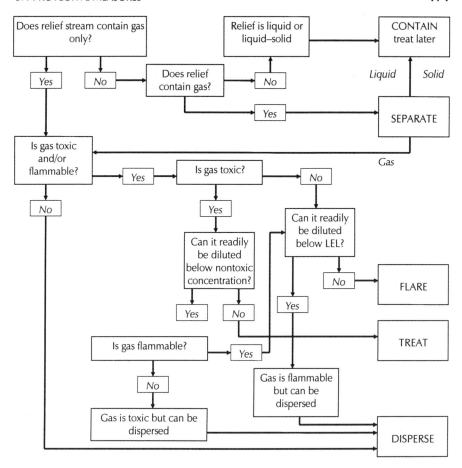

FIGURE 3.26. Decision Tree for Relief Disposal

An overview [230] and a decision tree [231] have been presented. Approaches to evaluate the venting possibilities are indicated in Figure 3.26. The following steps must be taken for the design of the relief system:

- establishing emission limits of possible discharges, either from the reaction system itself or from any vent collection system,
- identifying all potential emission sources,
- investigating possible process modifications,
- defining the control problem, and
- selecting a control system.

The sizing of the relief device requires data such as the physical form of the material likely to be released, the possible flow rate, the particle size distribution, variation of flow and concentrations, physical properties (e.g., density and solubility), chemical properties (e.g., corrosivity and toxicity), and temperature and pressure of the vented material.

Any stream containing liquid or solid or both should not be vented directly to the environment, which means essentially any relief vent stream since it is unusual that the stream would be fully as a gas. The liquid and/or solid should be collected and contained for later treatment. Typical collection and separation methods include knockout drums and cyclones. If the remaining gas stream contains toxic or flammable substances. it should also be treated or flared.

3.4.3.4 Dispersion, Flaring, Scrubbing, and Containment

An example of an overpressure protection system designed to reduce emissions to the atmosphere and at the same time provide adequate protection to the equipment has been described [234]. The equipment indicated is used for the manufacture of ethylene-vinyl acetate-vinyl chloride polymer emulsions. The design pressures are up to 100 bar.

The protection system consists of a combination of operator intervention alarms and automatic systems with bursting disc protected relief valves. This system gives a high level of integrity to the plant while allowing control of any over-pressurization situation. Downstream vent tanks are used to prevent liquid discharge. Flammable gases are vented through a high level stack with inert gas dilution.

If flammable gases are likely to be relieved to the atmosphere, the relief stream either cannot reach a source of ignition or, if ignited, it must burn harmlessly. Dilution of the relief stream to a point below the lower flammability limit is highly recommended. A review paper [231] provides details on solving this general problem.

In large installations, such as oil refineries and storage and shipping facilities, flare relief systems are usually provided as the methods of disposal for liquid and gaseous effluents from both normal and abnormal operations. A general discussion of the issues has been presented [235]. Combustion efficiencies can be well above 99%. Emergency flares must be able to handle difficult fuels over an enormous range of properties and concentrations and yet minimize pollution and noise. They must have reliable pilot burners and flame holders, and must give a long service life in rugged environments. In some cases, flame arresters are prescribed in order to prevent flash-backs which could cause tank and vessel explosions.

For the absorption of soluble gases or particulates from the gas streams, scrubbers can be used. These are generally simple packed columns, spray towers, or Venturi units. Design methods are well established [236]. Attention must be paid to the prevention of sprays, mists, and carryover when wet scrubbers are used.

If the effluent from the relief is highly toxic, containment or dilution is necessary. Containment systems range from relatively small drums used for

3.4 PROTECTIVE MEASURES

abnormal overfill situations to very sophisticated systems [237] for highly toxic materials. Total containment may involve designing the original equipment with sufficient strength to withstand any foreseen temperature and pressure rise, taking into account the possibility of a vacuum collapse as well.

3.4.3.5 Venting

Although venting is the subject of another AIChE/CCPS book [238], a brief discussion is presented here because venting is an essential part of safe reactor design. The physical form of the vented products may be gas or liquid, or mixtures of gas and liquid (two-phase flow). Mixtures containing solids may also be vented.

Experimental and calculation methods for gas and liquid venting are readily available. Two-phase flow used to be handled by overdesign of the venting area. Now, investigations into two-phase flow venting have led to considerably more insight into this situation and resulting phenomenon. References for vent system designs and applications include [191, 211, 233, 239, 240, 241, 242, 243, 244, 245].

Special packages have been developed for obtaining experimental information for designing vent systems, such as the VSP and ORSST discussed previously in Chapter 3. Calculation methods and software are available with these packages to design vent systems appropriately.

A specific problem can be approached in two ways : (1) use the design charts presented by First and Huff [232]. and (2) use the VSP package and calculation methods [195].

The design charts presented in [232] cover a range of vapor–liquid ratios from saturated liquid to saturated vapor. Homogeneous turbulent flows up to and including the maximum flow ("choked" value) conditions are considered. The method is not recommended for use at absolute temperatures above about 90% of the thermodynamic critical value, or at absolute pressures above about 50% of the thermodynamic critical value. These charts are most useful for single component and narrow-boiling multicomponent systems. Good results are obtained if an adiabatic program is available for tracking composition changes.

The use of the VSP equipment to obtain experimental data followed by appropriate calculations [195] is another approach to the vent system design issue.

4
MANAGEMENT OF CHEMICAL PROCESS SAFETY

4.1 HAZARD IDENTIFICATION AND QUANTIFICATION

The scope of this book includes several aspects of safe process design and operation, such as the choice of reactor type, safe operating conditions, and the selection of protective systems, primarily related to chemical reactivity. However, even in a process plant where these aspects have been carefully considered and thoroughly applied, there are still numerous events that can occur and can lead to hazardous incidents. Examples of such events are:

- leaking of a hazardous substance caused by corrosion,
- failures of the process control instrumentation,
- loss of power, cooling, electricity, or pressure, and
- human error (mischarging, opening of wrong valves, and so forth).

It is difficult to identify all of the possible events and their consequences in a complex chemical processing plant without the application of systematic procedures and proper management techniques. Several hazard evaluation procedures have been developed. Most of these procedures are described in other AIChE/CCPS publications such as *Guidelines for Hazard Evaluation Procedures* [2, 3] and *Guidelines for Quantitative Risk Analysis* [4]. Other publications on hazard evaluation techniques include [246, 247].

A few procedures that are used for identifying and evaluating potential hazards are summarized in Section 4.2. These methods are useful in many practical applications.

Identified hazards can either be reduced at the root (e.g., reduction of inventory) or by introduction of protective systems (e.g., automatic emergency shutdown). The increased understanding of the chemistry and the process that is obtained from hazard evaluation procedures provides guidance for many other elements of process safety management as well (e.g., procedural instructions, emergency strategies, personnel training, and preventive maintenance).

The evaluation of hazards in a process starts at the initial screening of the process parameters. Several technical issues that will need more attention will arise from this screening procedure. Hazard evaluation procedures, however, are no substitutes for engineering codes of practice and for design standards, but are used as supplementary ideas and concepts. A prerequisite for any process hazard evaluation is a full knowledge of the chemistry of the process (including potential unwanted side reactions) with supporting data.

The design of a process is most flexible in the early stages of development. The findings of hazard screening in this stage, although of a more general nature, indeed have the advantage that changes can be implemented at relatively low cost. Additional safety hazard studies can be carried out in a later stage of process development to ensure that the detailed engineering is consistent with the overall safety concepts already established. More detailed information on the use of hazard evaluation in different stages of the process design has been presented [248–250].

4.2 HAZARD EVALUATION PROCEDURES

Nine commonly used hazard evaluation procedures are highlighted in the short summary statements that follow. These procedures can be applied not only to the design of a new plant but also can be used to review safety conditions of existing plants, particularly regarding an update of current standards and operations.

1. Process System Checklist—
This procedure is based on the use of checklists which are applied to every stage of process design and operation to ensure compliance with standards, codes, good engineering practices, and well defined operating procedures. In this way, prior experiences can be implemented and used to prevent recurrence of incidents that may have happened in the past. Examples of checklists can be found in [2, 3, 251].

2. Dow and Mond Hazard Indexes—
The Dow and Mond Indexes provide a relative ranking of the hazards and risks in a chemical process plant. This is accomplished by assigning penalties and credits based on plant features such as the presence of hazardous materials and the safety devices which can mitigate any hazardous effects. Penalties and credits are then combined into a single hazard index for the process unit in question.

The indexes can be used to identify those units which require the highest priority for attention from a viewpoint of safety review. They are also useful in designing layout and spacing of equipment in process plants to avoid domino effects following an incident [2, 3, 252–254].

4.2 HAZARD EVALUATION PROCEDURES

3. *Preliminary Hazard Analysis (PHA)*—
PHA focuses on the hazardous materials and major plant elements in the process plant to provide a cost-effective hazard identification [2, 3]. It is intended for use in the early design stage and it can be very useful in site selection. It also provides early guidance to plant designers in considerations for reducing or eliminating potential hazards.

4. *What-if Analysis*—
A What-if analysis is used to assess consequences of deviations from normal operating conditions by asking "what if...?" questions. This approach is often used for reviewing plant or process modifications. Because the procedure is less structured than some other methods, e.g, HAZOP, care should be taken to identify also the less obvious hazards [2, 3, 255].

5. *Failure Modes, Effects, and Criticality Analysis (FMECA)*—
In the FMECA procedure [2, 3, 256], an exhaustive list of the equipment is first made. Every item on the list is then reviewed for possible ways in which it can fail (the failure modes are open, closed, leaks, plugged, on, off, etc.). The effects of each failure mode are then recorded and a criticality ranking of every item of equipment is calculated. A limitation of this procedure is that combinations of failures which may cause an incident are not really identified. Failure modes and effects analysis (FMEA) is the same procedure without the criticality analysis.

6. *Hazard Operability Study (HAZOP)*—
HAZOP is a systematic method to identify process deviations that could lead to incidents [2, 3, 251, 257, 258]. A multidisciplinary team works through piping and instrumentation diagrams or flowsheets applying certain guide words such as "more, " "less, " or "no" to process parameters such as flow, pressure, and temperature for every item in the diagram. All indicated deviations are checked for hazardous consequences and possibility of occurrence (a cause). The procedure can be used not only to identify hazards but also for operations and emergency control systems, and is especially useful when a new type of technology is involved in plant operations.

The advantage of HAZOP is the combination of the experiences of people from different disciplines which make the study very effective. Also, it enforces a certain discipline to minimize the possibility of overlooking any hazards.Several computer programs to assist in HAZOP studies are available [259, 260].

7. *Failure Logic Diagrams*—
A widely used method to study event sequences that can lead to incidents involves logic diagrams, an example of which is shown in Figure 4.1. Through the method illustrated in the figure, it is determined that a runaway reaction will happen only if both the cooling system fails and the reactor contents

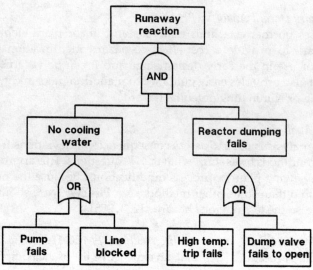

FIGURE 4.1. Example of a Fault Tree.

cannot be dumped into a catch tank. Moreover, it is shown that only four combinations can lead to a runaway.

Three hazard evaluation procedures using logic diagrams are: (1) fault-tree analysis (FTA), (2) event-tree analysis (ETA), and (3) cause–consequence analysis (CCA). Appropriate references are [2, 3, 251, 261].

8. *Human Error Analysis*—
Human Error Analysis is used to identify those conditions at which errors made by personnel are most likely to occur. It is often possible to reduce the likelihood of such errors by considering human factors in the design [262–264]. A few examples are presented below to illustrate this technique:

- Proper identification of equipment and a logical control panel arrangement reduce the chance of confusion and mistakes.
- If a simple action can lead to more serious consequences than is obvious, additional safeguards/barriers must be built in.
- Safety systems, including software, should be protected from unauthorized and/or accidental changes [265].
- As the level of automation increases, the operators assigned must be carefully considered and selected; operators should neither be left the simple tasks that cannot be automated, nor should their tasks become annoying ones; thus, the control system must provide the operators with sufficient information about the condition of the process, and they should have adequate facilities to interact with abnormal situations.

4.2 HAZARD EVALUATION PROCEDURES

9. *Quantitative Risk Assessment—*
The procedures described so far are generally of a qualitative nature, although the likelihood of events can be predicted using the failure logic diagram technique. A more quantitative method, such as quantitative risk assessment (QRA), also known as probabilistic risk assessment (PRA), may be required for a facility or a unit with a high hazard potential. In several cases, large scale studies have been performed to assess the risk an existing or planned industrial activity presents to the public [266, 267]. On a smaller scale, QRA can be a practical tool to reveal the most critical parts of a process and to determine which one of several design alternatives is the most effective from the viewpoint of safety. A full quantitative risk assessment consists of five steps: (1) hazard identification, (2) probability or frequency analysis, (3) consequence analysis, (4) calculations of risk level, and (5) assessment. Figure 4.2 illustrates a typical curve, known as the *F-n* curve, for frequency of incidents (*F*) and the number of fatalities (*n*) from calculations of a risk level. Example cases of the use of QRA as a decision aid are described in [270, 289]. More information can be found in [4, 261, 271].

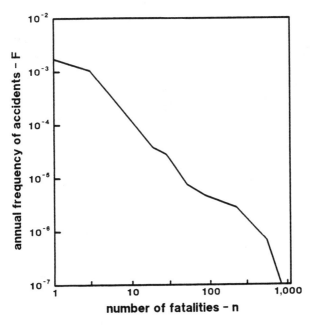

FIGURE 4.2. *F-n* Curve (Risk Curve).

4.3 CHEMICAL PROCESS SAFETY MANAGEMENT

In order to prevent incidents, a process plant must not only be well-designed, but also properly operated and maintained. To ensure that all safety aspects receive adequate priority, the commitment to safety from all levels of management is essential. In practice, conflicts of interest may arise between safety and other goals such as production demands and budgets. In these cases, the management attitude will be decisive. In reality, such a conflict of interests is only an apparent one because safety, efficiency, and product quality all depend on a reliable production facility with a low frequency of technical troubles and safety problems.

The Center for Chemical Process Safety (CCPS) of the American Institute of Chemical Engineers (AIChE) has identified twelve elements that must be part of any chemical process safety management program [5]. Application of these elements specifically to plant operations has also been defined [6]. Because of the critical importance of these twelve elements, they are listed here as follows for reference:

1. *Accountability: Objectives and Goals*—This element encompasses continuity of operations, continuity of systems (resources and funding), continuity of organizations, company expectations (vision and master plan), the quality process, control of expectations, alternative methods, management accessibility, and communications.
2. *Process Knowledge and Documentation*—The main features here are process definition and design criteria, process and equipment design, company memory (management information), documentation of risk management decisions, protective systems, normal and upset condtions, and chemical and occupational health hazards.
3. *Capital Project Review and Design Procedures*—For existing plants, expansions, and acquisitions, concerns must be addressed for appropriation request procedures, risk assessment for investment purposes, hazards review (including worst credible cases), siting (relative to risk management), plot plan, process design and review procedures, and project management procedures.
4. *Process Risk Management (Internal and Toll Operations)*—The key components are hazard identification (periodic process reviews of all operations inside the fence), risk assessment of operations, reduction of risk, residual risk management (in-plant emergency response and mitigation), process management during emergencies, and encouraging client and supplier companies to adopt similar risk management practices.
5. *Management of Change*—These items include change of technology, change of facility, organizational changes that may impact on process safety, variance procedures, temporary changes, and permanent changes.

6. *Process and Equipment Integrity*—Among the things to be considered here are reliability engineering, materials of construction, fabrication and inspection procedures, installation procedures, preventive maintainence, process, hardware and systems inspections and testing (pre-startup safety review), maintainence procedures, alarm and instrument management, and demolition procedures.
7. *Incident Investigation*—In assessing this element, consideration must be given to major incidents, near-miss reporting, follow-up and resolution, communications, incident recording, and third-party participation as needed.
8. *Training and Performance*—The key elements are definition of skills and knowledge, training programs (new employees, contractors, technical employees), design of operating and maintenance procedures, initial qualification assessment, on-going performance review and refresher training, instructor program, and records management.
9. *Human Factors*—Included here are human error assessment, operator/process and operator/equipment interfaces, and administrative controls versus hardware.
10. *Standards, Codes, and Laws*—Core concepts are internal standards, guidelines and practices (past history, flexible performance standards, amendments, and upgrades), and external standards, guidelines, and practices.
11. *Audits and Corrective Actions*—Items considered are process safety audits, maangement system audits and compliance reviews (criteria for internal/external reviews and auditors for external/internal reviews), and resolutions and close-out procedures.
12. *Enhancement of Process Safety Knowledge*—The level of performance in this area can be based on analysis of involvement in internal and external research, including CCPS programs and professional and trade association programs (both domestic and international), improved prredictive systems, such as toxicological data and trend information on maintenance failures, and a process safety reference library.

4.4 FUTURE TRENDS

As concluding remarks for this Guidelines book on chemical reactivity concerns, a few notes on future trends and issues regarding chemical process saftey are presented as follows:

- The changing societal, governmental, and industrial perspectives of risk will require that more detailed attention be placed on process safety factors in design and operation of chemical processing plants.

- Various government agencies will become partners in process safety issues, for example, the Occupational Safety and Health Administration in the United States [272] and the competent authorities in the European Union countries [273].
- Community groups near process plants will play an increasingly important role in assuring that catastrophic events do not occur.
- Environmental releases from incidents will have considerably less acceptance by government and community groups.
- Plant management personnel will have significantly greater roles in working with the community, for example, under the Responsible Care® program, of the Chemical Manufacturers Association [274–276];
- The manufacture and processing of chemical substances will involve increasingly complex technologies.
- Critical safety devices will be substatially improved with better sensors, more measurements, new technologies, and more reliability for critical and shutdowns;
- The need for the use of safety procedures by contract personnel will require additonal training and higher priorities in procedural reviews.
- Improved data logging and recall capabilities with new instrumentation will enhance practical knowledge of the processes;
- Plant simulators will provide better operator training.
- Just-in-time inventory systems will reduce the quantities of stored hazardous materials.
- Artificial intelligence concepts will be applied in assessing process measurements and controls.

Essentially all aspects of chemical process safety, including issues related to chemical reactivity, are dynamic.

REFERENCES

REFERENCES CITED

1. Kohlbrand, H. T., "The Relationship between Theory and Testing in the Evaluation of Reactive Chemical Hazards," in *Proceedings of the International Symposium on Prevention of Major Chemical Accidents*, p. 4. 69, Center for Chemical Process Safety/AIChE, New York, NY (1987).
2. *Guidelines for Hazard Evaluation Procedures*, Center for Chemical Process Safety/AIChE, New York, NY (1985).
3. *Guidelines for Hazard Evaluation Procedures (Second Edition) with Workbook*, Center for Chemical Process Safety/AIChE, New York, NY (1992).
4. *Guidelines for Chemical Process Quantitative Risk Analysis*, Center for Chemical Process Safety/AIChE, New York, NY (1989).
5. *Guidelines for Technical Management of Chemical Process Safety*, Center for Chemical Process Safety/AIChE, New York, NY (1988).
6. *Plant Guidelines for Technical Management of Chemical Process Safety*, Center for Chemical Process Safety/AIChE, New York, NY (1992).
7. *Guidelines for Auditing Process Safety Management Systems*, Center for Chemical Process Safety/AIChE, New York, NY (1993).
8. *Guidelines for Engineering Design for Process Safety*, Center for Chemical Process Safety/AIChE, New York, NY (1993).
9. West, A. S., "Chemical Reactivity Evaluation: The CCPS Program," *Process Safety Progress*, **12**, 34 (January 1993).
10. Yoshida, T., *Safety of Reactive Chemicals*, Elsevier, Amsterdam, The Netherlands (1987).
11. Ducros, M. and H. Sannier, "Application du Programme CHETAH a l'Etude de la Sensibilisation de Composes Oxygenes et a l'Estimation des Limites Interferieures d'Inflammabilite" ("Application of the CHETAH Program for the Study of the Sensibility of Oxygen Containing Mixtures and the Estimation of the Lower Flammability Limits"), *J. Haz. Mat.*, **19** (1988).
12. Stull, D. R., *Fundamentals of Fire and Explosion*, American Institute of Chemical Engineers Monograph Series #10, New York, NY (1977).

13. Coffee, R. D., "Chemical Stability," in *Safety and Accident Prevention in Chemical Operations*, H. H. Fawcett and W. S. Wood (editors), p. 305, John Wiley, New York, NY (1982).
14. *Guidelines for a Reactive Chemicals Program*, Dow Chemical Company, Midland, MI (1987).
15. Regenass, W., "Safe Operation of Exothermic Reactors," technical paper, Ciba-Geigy Limited, Basel, Switzerland (1984).
16. Gibson, N., "Hazard Evaluation and Process Design," in *Proceedings of Runaway Chemical Reaction Hazards Symposium*, IBC, London, England (1987).
17. Cronin, J. L., B. F. Noland, and J. A. Barton, "A Strategy for Thermal Hazard Assessment in Batch Chemical Manufacturing," *I. Chem. E. Symp. Series*, **102**, UMIST, Manchester, England (June 1987).
18. King, P. V., and A. H. Lasseigne, "Hazard Classification of Explosives for Transportation: Evaluation Test Methods, Hazard Evaluation, Research, and Engineering," Report No. TSA-20-72-5, prepared for U. S. Dept. of Transportation (1972) (available from NTIS, Springfield, VA).
19. Dale, C. B., "Hazard Classification of Explosives for Transportation—Non-solid Explosives Phase III," Naval Ordnance Station, Report No. DOT/RSPA/MTB-78/2 (1978).
20. Coffee, R. D., "Hazard Evaluation Testing," CEP Technical Manual, *Loss Prevention*, **3**, 18 (1969), American Institute of Chemical Engineers, New York, NY.
21. Zatka, A. V., "Application of Thermal Analysis in Screening for Chemical Process Hazards," *Thermochimica Acta*, **28**, 7 (1979).
22. United Nations, "Recommendations on the Transport of Dangerous Goods, Part I, Tests and Criteria for the Classificatioin of Explosive Substances and Articles", Report ST/SG/AC. 10/11/Rev 1, United Nations, New York, NY (1990).
23. Grewer, T., H. Klusacek, U. Loffler, R. L. Rogers, and J. Steinbach, "Determinations of the Characteristic Values for Evaluation of the Thermal Safety of Chemical Processes," *J. of Loss Prev. Process Ind.*, **2** (1989).
24. Meyer, R., *Explosives*, Verlag Chemie, New York, NY (1977).
25. Frurip, D. J., E. Freedman, and G. R. Hertel, "A New Release of the ASTM CHETAH Program for Hazard Evaluation: Versions for Mainframe and Personal Computer," in *Proceedings of the International Symposium on Runaway Reactions*, Center for Chemical Process Safety/AIChE, New York, NY (1989).
26. Davies, C. A., I. M. Kipnis, M. W. Chase, and D. N. Trewek, "The Thermochemical and Hazard Data of Chemicals: Estimation Using the ASTM CHETAH Program," ACS Symposium Series, **274**, 82 (1985), American Chemical Society, Washington, DC.
27. Seaton, W. H., E. Freedman, and D. N. Treweek, "CHETAH: The ASTM Chemical Thermodynamic and Energy Release Potential Evaluation Program," American Society for Testing and Materials, Philadelphia, PA (1974).
28. Grewer, Th., "Exotherme Zersetzung: Untersuchung der Charakteristischen Stoffeigenschaften" ("Exothermic Decomposition: Investigation of the Characteristic Properties"), VDI Verlag, Dusseldorf, Germany (1988).

29. Fierz, H., and G. Zwahlen, "Deflagration in Solids—A Special Hazard in Chemical Production, *Proceedings of the 6th Loss Prevention and Safety Promotion in the Process Industries Symposium*, Oslo, Norway (1989).
30. Gygax, R. W., "Scale-up Principles for Assessing Thermal Runaway Risks," *Chem. Eng. Prog.*, **86**, 53 (February 1990).
31. Bretherick, L., "Reactive Chemical Hazards: An Overview," *Proceedings of the International Symposium on Preventing Major Chemical Accidents*, p. 4. 1, Center for Chemical Process Safety/AIChE, New York, NY (1987).
32. Grewer, Th., "Unerwunschte Reaktionen Organischer Stoffe als Gefahren in Chemieanlagen" ("Hazards of Undesired Reactions of Organic Compounds in Chemical Plants"), *Chem. Ing. Tech.*, **51**, No. 10, 928 (1979).
33. Stull, D. R., "Prediction of Real Chemical Hazards," presented at 65th Annual Meeting of the American Institute of Chemical Engineers, New York, NY, 1973.
34. Seaton, W. H., "Group Contribution Method for Predicting the Potential of a Chemical Composition to Cause an Explosion," *Safety in the Chemical Laboratory*, **66**, No. 5, A137 (1989).
35. Bretherick, L., *Handbook of Reactive Chemical Hazards*, 4th Edition, Butterworth-Heinemann, Stoneham, MA (1990).
36. Leleu, J., serially in *Cahiers de Notes Documentaires (Record Book of Documented Notes)*, 69-319 to 100-421 (1972–1980).
37. Nagel, M. C. (ed.), "Peroxides Can Be Treacherous," *J. Chem. Educ.*, **61**, 250 (1984).
38. *CHAS Notes* **1**, No. 4, 1, American Chemical Society Division of Chemical Health and Safety, Washington, DC (1982).
39. McCloskey, C. . . M., "Safe Handling of Organic Peroxides: An Overview," *Plant/Operations Progress*, **8**, 185 (1989).
40. deGroot, J. J., Th. M. Groothuizen, and J. Verhoeff, "Safety Aspects of Organic Peroxides in Bulk Tanks," *Ind. Eng. Chem. Process Des. & Dev.*, **20**, 131 (1981).
41. Noller, D. C., S. J. Mazurowski, and G. F. Linden, *Ind. Eng. Chem.*, **56**, No. 12, 19 (1964).
42. Stull, D. R., "Linking Thermodynamics and Kinetics to Predict Real Chemical Hazards," CEP Technical Manual, *Loss Prevention*, **7**, 67 (1988), American Institute of Chemical Engineers, New York, NY.
43. Bretherick, L., "Chemicals, Reactions, Hazards," in *Safety & Loss Prevention in Chem. Oil Processing Ind. Symposium*, **D5** (1989), Singapore.
44. Stull, D. R., "Identifying Chemical Reaction Hazards," CEP Technical Manual, *Loss Prevention*, **4**, 16 (1970), American Institute of Chemical Engineers, New York, NY.
45. Lothrop, W. D. and G. R. Handrick, *Chem. Rev.*, **44**, 419 (1949).
46. Jain, S. R., "Energetics of Propellants, Fuels, and Explosives: A Chemical Valence Approach," *Propellants Explos. Pyrotech.*, **12** (1987).
47. Weast, R. C. (ed.), *Handbook of Chemistry and Physics*, 66th Edition, CRC Press, Inc., Boca Raton, FL (1985).
48. Stull, D. R., E. F. Westrum, and G. C. Sinke, *The Chemical Thermodynamics of Organic Compounds*, John Wiley, New York, NY (1969).

49. Benson, S. W., *The Foundation of Chemical Kinetics*, Robert E. Krieger Publ. Co., Malabar, FL (1982).
50. Craven, A. D., "A Simple Method of Estimating Exothermicity by Average Bond Energy Summation" in *Hazards from Pressure: Exothermic Reactions, Unstable Substances, Pressure Relief and Accidental Discharge, I. Chem. Eng. Symp. Series*, **102**, 97 (1987).
51. Anderson, J. W., G. H. Beyer, and K. M. Watson, *Natl. Pet. New Tech. Sec.*, **36** (1944).
52. Hougen, D. A., K. M. Watson, and R. A. Ragatz, *Chemical Process Principles*, 2nd Edition, Part II—*Thermodynamics*, John Wiley, New York, NY (1959).
53. Yoneda, Y., *CHEMOGRAM I—Computer Industrial Chemistry*, Maruzen Book Co., Tokyo, Japan (1972).
54. Martinez, M. R., "Estimation of Gas-Phase Thermodynamic Parameters," Quantum Chemistry Program Exchange, Program 244 (1974).
55. Sanderson, R. T., *Chemical Bond and Bond Energy*, Academic Press, New York, NY, and London, England (1971).
56. DIPPR Reports #802-5-83 and #802-11-87, Design Institute for Physical Property Research, American Institute of Chemical Engineers, New York, NY.
57. Treweek, D. N., J. R. Hoyland, C. A. Alexander, and W. M. Pardue, "Use of Simple Thermodynamic and Structural Parameters to Predict Self-Reactivity Hazard Ratings of Chemicals," *J. Haz. Mat.*, **1** (1975/1976).
58. Yoshida, T., M. Itoh, and K. Nagai, "A Preliminary Evaluation of Incompatibility and Energy Release," in *Proceedings of the 4th Symposium on Chemical Problems Connected with the Stability of Explosives*, p. 159, Sweden (1976).
59. Cowperthwaite, M. and W. H. Zwisler, "TIGER: Computer Program Documentation," Stanford Research Institute, Menlo Park, CA (1973).
60. Gordon, S., and B. J. McBride, "Chemical Program for Calculation of Complex Chemical Equilibrium Compositions, Rocket Performance, Incident and Reflected Shocks, and Chapman-Jouguet Detonations," NASA Lewis Research Center, National Aeronuatics and Space Administration, Washington, DC (1971).
61. National Fire Protection Association, "Manual of Hazardous Chemical Reactions," NFPA, 491M, Quincy, MA (1991).
62. Yoshida, T. (ed.), Tokyo Fire Department Compilation: *Handbook on Hypergolic Ignition*, Nikkan Kogyo Shinbunsha, Tokyo, Japan (1980).
63. Leach, J. T. and J. M. Stephens, "Hazards of Inadvertent Mixing of Chemicals Used in the Bachmann Processes for Manufacturing the Military Explosives RDX and HMX," *J. Haz. Mat.*, **4**, 271–281 (1981).
64. Murali, B. K., V. Mohan, K. Ganesan, and R. Bhujanga, "Hazard Characteristics of Isoboride Dinitrate–Lactose Mixtures," *J. Haz. Mat.*, **3**, 77–182 (1979).
65. Treweek, D. N., C. A. Alexander, J. R. Hoyland, A. F. Pentiman, Jr., and W. M. Pardue, "Estimation of Chemical Incompatibility (Other-Chemical Reactivity) by Computer," *Ohio J. Sci.*, **80**, No. 4, 160–166 (1979).
66. Jackson, H. L., W. B. McCormack, C. S. Rondestvedt, K. C. Smeltz, and I. E. Viele, "Control of Peroxidizable Compounds," *J. Chem. Educ.*, **47**, No. 3 (1970).
67. Anon., "Oxidizing Agents—Hazards and Precautions," *Fire Prevention*, **127**, 30.

68. Statesir, W. A., "Explosive Reactivity of Organics and Chlorine," *Chem. Eng. Prog.,* **69**, No. 4, 114 (1973).
69. Cutler, D. P., "Reactions between Halogenated Hydrocarbons and Metals—A Literature Review," *J. Haz. Mat.,* **17**, 99 (1987).
70. Lombardi, G., "For Better Thermal Analysis," *International Conference for Thermal Analysis, 2nd Edition,* Rome, Italy (1980).
71. McNaughton, J. L. and C. T. Mortimer, *IRS: Physical Chemistry Series 2,* Volume 10, Butterworth, London, England (1975).
72. Wendtlandt, W., *Thermal Methods of Analysis,* 2nd Edition, John Wiley, New York, NY (1974).
73. Wunderlich, B., *Differential Thermal Analysis, Techniques and Chemistry,* Volume I, Part V (1971).
74. Mackenzie, R. C., *Differential Thermal Analysis,* Parts I and II, Academic Press, New York, NY (1972).
75. Yoshida, T., M. Itoh, K. Tohyama, and M. Tamura, "The Correlation of DSC Data with Other Test Data," *Chemical Problems Connected with the Stability of Explosives,* **5**, Part I, 79 (1979).
76. Zeman, S., "The Relationship between Differential Thermal Analysis Data and the Detonation Characteristics of Polynitroaromatic Compounds," *Thermochimica Acta,* **41** (1980).
77. Fenlon, W. J., "A Comparision of ARC and Other Thermal Stability Test Methods," *Pkant/Operations Progress,* **3**, No. 4, 197 (1984).
78. Horowitz, M. H., and G. Metzger, *Anal. Chem.,* **35**, 1464 (1963).
79. Cronin, J. L., and P. F. Nolan, "The Comparative Sensitivity of Test Methods for Determining Initial Exotherm Temperatures in Thermal Decompositions of Single Substances," *J. Haz. Mat.,* **14** (1987).
80. Seyler, R. J., "Applications of Pressure DTA (DSC) to Thermal Hazard Evaluation," *Thermochimica Acta,* **39** (1980).
81. Hentze, G., "Safety Tests for Production in Chemical Industry using Differential Thermal Analysis," *Thermochimica Acta,* **72** (1984).
82. Mix, K. K., "The Use of Advanced DTA Methods for the Evaluation of Thermal Instability, Hazard Evaluation, and Process Design," in *Proceedings Runaway Chemical Reaction Hazards Symposium,* IBC, London, England (1987).
83. Ozawa, T., "Kinetic Analysis of Derivative Curves in Thermal Analysis," *J. of Therm. Anal.,* **2**, 301 (1979).
84. *Annual Book of ASTM Standards,* ASTM Methods E-537-84 and E-472-79, Volume 14. 02, American Society for Testing and Materials, Philadelphia, PA (1994).
85. *Annual Book of ASTM Standards,* ASTM Methods E-967-83, E-474-80, and E-968-83, Volume 14. 02, American Society for Testing and Materials, Philadelphia, PA (1994).
86. OECD Guideline for Testing of Chemicals, "Screening Test for Thermal Stability and Stability in Air," No. 113, Organization for Economic Cooperation and Development, Paris, France (1981).

87. *Guidelines for Chemical Reaction Hazard Evaluation*, The Association of the British Pharmaceutical Industry, London, England (1989).
88. O'Neill, M. J., "A Study of Exothermic Reactions by Differential Scanning Calorimetry," *Anal. Chem., Analyst*, **99** (1974).
89. Gibson, N., R. L. Rogers, and T. K. Wright, "Chemical Reaction Hazard," *I. Chem. Symp. Series*, **102** (1987).
90. Hofelich, T. C. and R. C. Thomas, "The Use/Misuse of 100-Degree Rule in the Interpretation of Thermal Hazard Tests," in *Proceedings of the International Symposium on Runaway Reactions*, Center for Chemical Process Safety/AIChE, New York, NY (1989).
91. Duswalt, A. A., "The Practice of Obtaining Kinetic Data by Differential Scanning Calorimetry," *Thermochimica Acta*, **8**, 56 (1974).
92. Skvara, F. and V. Satava, "Kinetic Data from DTA Measurements," *J. of Thermal Analysis*, **2**, 325 (1970).
93. Merzhanov, A. G., V. V. Barzykin, A. S. Shteinberg, and V. T. Gontkovskaya, "Methodical Principles of Studying Chemical Reaction Kinetics under Conditions of Programmed Heating," *Thermochimica Acta*, **21**, 301 (1977).
94. Freeman, E. S., and B. Carroll, "The Application of Thermoanalytical Techniques to Reaction Kinetics," *J. Phys. Chem.*, **62**, 394 (1958).
95. Rogers, R. N., and L. C. Smith, "The Application of Scanning Calorimetry to the Study of Chemical Kinetics," *Thermochimica Acta*, **1** (1970).
96. Gyulai, G. and E. J. Greenhow, *J. of Thermal Analysis*, **6**, 279 (1974).
97. Watson, E. S., M. J. O'Neill, J. Justin, and N. Brenner, *Anal. Chem.*, **36**,1233 (1964).
98. Borchardt, H. J. and F. Daniels, "The Application of Differential Thermal Analysis to the Study of Reaction Kinetics," *J. Amer. Chem. Soc.*, **79**, 41 (1957).
99. Kissinger, H. E., "Reaction Kinetics in Differential Thermal Analysis," *Anal. Chem.*, **29**, 1702 (1957).
100. Kissinger, H. E., *J. Nat. Bur. Stand.*, **57**, 217 (1956).
101. Coats, A. W. and J. P. Redfern, "Kinetic Parameters from Thermogravimetric Data," *Nature*, **210**, 68 (1964).
102. "A System for Low Cost Thermal Hazards Screening," RADEX-solo, SYSTAG Information Documentation (1990).
103. Hakl, J., "Advanced SEDEX (Sensitive Detector of Exothermic Processes)," *Thermochimici Acta*, **80**, 109 (1984).
104. Hakl, J., "Sensitive Detector of Exothermic Processes (SEDEX)," *Thermochimici Acta*, **38**, 253 (1980).
105. Kupr, T., "Sicherheitstechnische Beurteilung van Chemieverfahren, Notige Grossen und Parameter" ("Safety Technical Assessment of Chemical Processes, Necessary Quantities and Parameters"), ZSL-Labor, Sandoz AG, Basel, Switzerland (1989).
106. Hub, L., "Two Calorimetric Methods for the Investigation of Dangerous Reactions," *Chemical Process Hazards VI, I. Chem. Eng. Symp. Series*, **49** (1977).
107. *Annual Book of ASTM Standards*, ASTM Method E-771-81, Volume 14. 02, American Society for Testing and Materials, Philadelphia, PA (1994).

108. Hub, L., "Universal System for Investigation of Thermal Hazards," *Thermochimica Acta*, **85** (1985).
109. Porter, R. S., and J. F. Johnson, *Analytical Calorimetry*, Plenum Press, New York, NY (1968).
110. Reid, D. S., "Differential Microcalorimeters," *J. of Physics, E: Scientific Instruments*, **9** (1976).
111. Wright, T. K., and C. W. Butterworth, "Isothermal Heat Flow Calorimeter," *I. Chem. Eng. Symp. Series*, **102** (1987).
112. Wadso, I., "Chemical Problems Connected with the Stability of Explosives," *Microcalorimetry*, **5**, Part I (1979).
113. Calvet, E. and H. S. Prat, *Recent Progress in Microcalorimetry*, Pergamon Press, London, England (1963).
114. "UN Recommendations on the Transport of Dangerous Goods, Part II, Test Methods for Determining the Self-Accelerating Decomposition Temperature (SADT) of Organic Peroxides and Other Thermally Unstable Substances," ST/SG/AC. 10/11/Rev. 1, United Nations, New York, NY (1990).
115. Van Geel, J. L. C., "Self-Ignition Hazard of Nitrate Ester Propellants," Thesis, Technical University of Delft, The Netherlands (1969).
116. Van Geel, J. L. C., "Safe Radius of Heat Generating Substances," *Ind. Eng. Chem.*, **58**, No. 1, 24 (1968).
117. Waters, D. N. and J. L. Paddy, "Equations for Isothermal Differential Scanning Calorimetry," *Anal. Chem.*, **60** (1988).
118. Grewer,Th., "Dewar Test Methods for Exothermic Reactions," in *Proceedings Runaway Chemical Reaction Hazards Symposium*, IBC, London, England (1987).
119. Lemke, D., "Waermestauverfahren und Adiabatische Lagerung zur Sicherheitslichen Beurteilung Thermisch Instabiler Stoffe" ("Heat Accumulation and Adiabatic Storage Test for the Classification of Thermally Unstable Compunds"), Colloquy on the Prevention of Accidents at Work and Occupational Diseases in Chemical Industries (IVSS), Heidelberg, Germany (1976).
120. Groothuizen, Th. M., J. W. Hartgerink, and H. J. Pasman, "Phenomenology, Test Methods, and Case Histories of Explosions in Liquids and Solids," *Loss Prev. Safety Prom. Process Ind.*, 239 (1974).
121. Pasman, H. J., Th. M. Groothuizen, and C. H. Vermeulen, *Explosivstoffe*, **17**, 151 (1969).
122. Wright, T. K. and R. L. Rogers, "Adiabatic Dewar Chemistry," in *Hazards in the Process Industries: Hazards IX, I. Chem. Eng. Symp. Series*, **97**, 121 (1986).
123. Townsend, D. and J. C. Tou, "Thermal Hazard Evaluation by an Accelerating Rate Calorimeter," *Thermochimici Acta*, **37**, 1 (1980).
124. Townsend, D., "Accelerating Rate Calorimetry," *I. Chem. Eng. Symp. Series*, **68** (1981).
125. Smith, D. et al., "Accelerating Rate Calorimetry," *Am. Lab.*, **12**, No. 6 (1980).
126. Tou, J. C. and L. F. Whiting, "The Thermokinetic Performance of an Accelerating Rate Calorimeter," *Thermochimica Acta*, **48** (1981).

127. De Haven, E. S., "Approximate Hazard Ratings and Venting Requirements from CSI-ARC Data", *Plant/Operations Progress*, **2**, No. 1 (1983).
128. Snee, T. J., "Incident Investigation and Hazard Evaluation Using Differential Scanning Calorimetry and Accelerating Rate Calorimetry," *J. Occupational Accidents*, **8** (1987).
129. Fenlon, W. J., "Calorimetric Methods Used in the Assessment of Thermally Unstable or Reactive Materials," in *Proceedings of the International Symposium on Prevention of Major Chemical Accidents*, Center for Chemical Process Safety/AIChE, New York, NY (1987).
130. Coates, C. F., "The ARC in Chemical Hazard Evaluation," *Thermochimica Acta*, **85** (1985).
131. Kohlbrand, H. T., D. I. Townsend, and M. Brierly, "Application of the ARC in an Industrial Reactive Chemicals Program," in *Proceedings of the Runaway Reaction Hazards Symposium*, IBC, London, England (1987).
132. Townsend, D. I., "Hazard Evaluation of Self-Accelerating Reactions," *Chem. Eng. Prog.*, **73**, 80 (1977).
133. Bowes, P. C., *Self-Heating: Evaluating and Controlling the Hazards*, 1st edition, Elsevier, Amsterdam, The Netherlands (1984).
134. United Nations Recommendations on the Transport of Dangerous Goods, ST/SG/AC. 10/1, Rev. 6, United Nations, New York, NY (1989).
135. 92/32/EEC, Seventh Amendment to EEC Directive 67/548/EEC, "Directive on Classification, Packaging, and Labelling of Dangerous Substances," European Union Commission, Brussels, Belgium (June 5, 1992).
136. Treumann, H., G. Kruger, N. Pfeil, and S. von Zahn-Ullmann, "Sicherheitstechnische Kenndaten und Gefahrzahlen Binarer Mischungen aus Oxidierenden und Verbrennlichen Substanzen" ("Safety Data and Hazard Rating of Binary Mixtures of Oxidizers and Flammable Substances"), Forschungbericht 142, Bundesansalt für Materialfirschung und -Prüfung (BAM), Berling, Germany (1987).
137. Gibson, N., D. J. Harper, and R. L. Rogers, "Hazards from Exothermic Decomposition of Powders," in *Proceedings Runaway Chemical Reaction Hazards Symposium*, IBC, London, England (1987).
138. Schell, J. M. and A. H. Heemskerk, "On the Spontaneous Ignition of Material Deposited in Layers in Dryers and Filters," *Proceedings Eurotherm Seminar*, **14**, Louvain-la-Neuve, Belgium (1990).
139. Koenen, H., K. H. Ide, and K. H. Swart, "Sicherheitstechnische Kenndaten Explosionsfahigen Stoffe" (Safety Technical Characteristic Properties of Explosive Substances"), *Explosivestoffe*, **9**, No. 4, 30 (1961).
140. Mason, M. C. and E. G. Aiken, "Methods of Evaluating Explosives and Hazardous Materials," U. S. Bureau of Mines IC 8541 (1972).
141. Conner, J., "Explosion Risks of Unstable Substances, Test Methods Employed by EM 2 (Home Office) RARDE," in *Proceedings First International Loss Prevention Symposium*, The Hague, The Netherlands (1974).
142 Bakham, N. N. and A. F. Belyaev, "Combustion of Heterogeneous Condensed Systems," Translation edited by Ministry of Technology, London, England (1967).

143. United Nations Recommendations on the Transport of Dangerous Goods, Part III, Tests and Criteria for the Classification of Organic Peroxides, ST/SG/AC. 10/11, Rev. 1, United Nations, New York, NY (1990).
144. Merzhanov, A. G., *Combustion and Flame*, **13** (1969).
145. United Nations Recommendations on the Transport of Dangerous Goods, Part IV, Tests and Criteria for the Classification of Miscellaneous Dangerous Goods, Annex 3, page 37, ST/SG/AC. 10/11, Rev. 1, United Nations, New York, NY (1990).
146. Noronha, J. A., M. R. Juba, C. S. Brown, Jr., and J. G. Schmitt, "The Kinetics of Runaway Reactions," CEP Technical Manual, *Loss Prevention*, **13**, 50 (1980), American Institute of Chemical Engineers, New York, NY.
147. Grewer, T. and O. Klais, "Pressure Rise during Homogeneous Decomposition and Deflagration," *I. Chem. Eng. Symposium Series*, **102** (1987).
148. Verhoeff, J. and A. H. Heemskerk, "Gas Production and Heat Generation Data Concerning the Thermal Runaway of Liquid Reaction Systems," *I. Chem. Eng. Symposium Series*, **85** (1984).
149. Groothuizen, Th. M., J. Verhoeff, and J. J. De Groot, "Determination of the Effect of Thermal Explosion of Organic Peroxides," *J. Haz. Mat.*, **2** (1978).
150. Turner, B. C., "The Time/Pressure Test, Part I, Design and Experimental Method," EM2 Branch, RARDE, London, England (1973).
151. Hildenbrand, D. L., A. G. Whittaker, and C. B. Euston, *J. Phys. Chem.*, **58** (1954).
152. Pesetsky, B., J. N. Cawse, and W. T. Vyn, "Liquid Phase Decomposition of Ethylene Oxide," CEP Technical Manual, *Loss Prevention*, **13**, 123 (1980), American Institute of Chemical Engineers, New York, NY.
153. Feath, G. M., "High-Pressure Liquid-Monopropellant Strand Combustion," *Combustion and Flame*, **18** (1972).
154. Groothuizen, Th. M., and H. J. Pasman, "Explosions in Liquids and Solids," CEP Technical Manual, *Loss Prevention*, **9**, 91 (1975), American Institute of Chemical Engineers, New York, NY.
155. Annual Book of ASTM Standards, ASTM Method E-680-79, Volume 14. 02, American Society for Testing and Materials, Philadelphia, PA (1994).
156. Koenen, H., and K. H. Ide, "Uber die Prüfung Explosiever Stoffe II, Ermittlung der Reibempfindlichkeit van Zund- (Initial-) Sprengstoffen und Andere Sehr Reibempfindlichkeit Explosiven Stoffen" ("On the Testing of Explosive Substances II, Measurements of the Friction Sensitivity of Primary Explosives and Other Very Friction Sensitive Explosive Substances"), *Explosivestoffe*, **1** (1956).
157. Kuchta, J. M., A. L. Furno, and A. C. Imhof, "Classification Test Methods foir Oxidizing Materials," U. S. Bureau of Mines, Report of Investigation 7594 (1972).
158. Hasegawa, K., K. Kazutomo, H. Yoshihisa, and N. Hidefumi, "Characteristics of Conical Pile-Type Burning Test Methods for Determining the Potential Hazards of Oxidizing Materials," *J. Loss Prev. Process Ind.*, **2**, 135 (1989).
159. *Guidelines for Evaluating the Chacteristics of Vapor Cloud Explosions, Flash Fires, and BLEVES*, Center for Chemical Process Safety/AIChE, New York, NY (1994).

160. Regenass, W., "The Control of Exothermic Reactions," *I. Chem. Eng. Symposium Series*, **85**, 1 (1984).
161. Gygax, R., "Explicit and Implicit Use of Scale-Up Principles for the Assessment of Thermal Runaway Risks in Chemical Production," in *Proceedings of the International Symposium on Runaway Reactions*, Center for Chemical Process Safety/AIChE, New York, NY (1989).
162. Frank-Kamenetskii, D. A., *Zhur. Fiz. Khim.*, **13**, 378 (1939).
163. Duswalt, A. A., private communication.
164. ASTM E-698, "Test Method for Arrhenius Kinetic Constants for Thermally Unstable Materials," revised 1988, American Society for Testing and Materials, Philadelphia, PA.
165. Semenov, N. N., *Zhur. Physik.*, **48**, 571 (1928).
166. ASTM E-1231, "Standard Practice for Calculation of Hazard Potential Figures-of-Merit for Thermally Unstable Materials," 1988, American Society for Testing and Materials, Philadelphia, PA.
167. Westerterp, K. R., W. P. M. van Swaaij, and A. A. C. M. Beenackers, *Chemical Reactor Design and Operation*, John Wiley, New York, NY (1984).
168. Levenspiel, O., *Chemical Reaction Engineering*, 2nd Edition, John Wiley, New York, NY (1972).
169. Aris, R., *Elementary Chemical Reactor Analysis*, Prentice-Hall, Englewood Cliffs, NJ (1969).
170. Balzhiser, R. E., M. R. Samuels, and J. D. Eliasser, *Chemical Engineering Thermodynamics*, Prentice-Hall, Englewood Cliffs, NJ (1972).
171. Bird, R. B., W. E. Stewart, and E. N. Lightfoot, *Transport Phenomena*, John Wiley, New York, NY (1960).
172. Heemskerk, A. H., "Dynamical Behavior of Liquid-Phase Reaction System," Thesis, University of Amsterdam, The Netherlands (1980).
173. Boynton, D. E., W. B. Nichols, and H. W. Spurlin, "How to Tame Dangerous Reactions (Control of Exothermic Reactions)," *Ind. Eng. Chem.*, **51**, 489 (1959).
174. Cummings, G. H., and A. S. West, "Heat Transfer Data for Kettles with Jackets and Coils," *Ind. Eng. Chem.*, **42**, 2303 (1950).
175. Steensma, M., "Safe Operation of a Cooled Semi-Batch Reactor, Slow Liquid–Liquid Reactions," *Chem. Eng. Sci.*, **43**, 2125 (1988).
176. Coulson, J. M., and J. F. Richardson, *Chemical Engineering*, Volumes I, II, and III, Pergamon Press, Oxford, England (1977).
177. Steinbach, J., "Untersuchung zur Thermischen Sicherheit des Indirekt Gekuehlten Semi-Batch Reaktors" ("Investigation into the Thermal Safety of Indirectly Cooled Semi-Batch Reactors"), Thesis, University of Berlin, Germany (1985).
178. Steinbach, J., "Grenzen der Interpretierbarkeit Thermokinetischer Messungen" ("Limits to the Evaluation of Thermokinetic Measurements"), DECHEMA Tagung, Bad Soden, Germany (1990).
179. Bourne, J. R., F. Brogli, F. Hoch, and W. Regenass, "Heat Transfer from Exothermically Reacting Fluids in Vertical Unstirred Vessels—I. Temperature and Fluid Fields," *Chem. Eng. Sci.*, **42**, 2183 (1987).

180. Fletcher, F., "Heat Transfer Coefficients for Stirred Batch Reactor Design," *The Chemical Engineer*, **33** (1987).
181. Lees, F. P., *Loss Prevention in the Process Industries, Hazard Identification, Assessment, and Control*, Volume I, London, England (1980).
182. Hoppe, T., and R. Bruderer, "Thermal Analytical Investigation into a Runaway Reaction," in *Proceedings of the 6th Annual Symposium on Loss Prevention and Safety Promotion in the Process Industries*, p. 37, Oslo, Norway (1989).
183. Regenass, W., "Thermische Methoden zur Bestimmung der Makrokinetik" ("Thermal Methods to Determine Macro-Kinetics"), *Chimia*, **37**, 430 (1983).
184. Technical Equipment Documentation about RC1, Mettler Toledo, Hightstown, NJ (1994).
185. Technical Equipment Documentation about Contalab, Mettler Toledo, Hightstown, NJ (1992).
186. Poersch-Panke, H. G., A. Avela, and K. H. Reichest, "Ein Reaktionskalorimeter zur Untersuchung van Polymerisations" ("A Reaction Calorimeter for the Investigation of Polymerizations"), *Chem. Ing. Tech.*, **61**, 608 (1989).
187. Lox, E., F. Coenen, R. Vermeulen, and F. Froment, "A Versatile Bench-Scale Unit for Kinetics Studies of Catalytic Reactions," *Ind. Eng. Chem. Res.*, **27**, 576 (1988).
188. Technical Equipment Documentation from ThermoMetric AB, Sweden (1990).
189. Hub, L. and T. Kupr, "Heat Balance Calorimetry and Automation of the Testing Procedure," in *Proceedings of the Runaway Chemical Reaction Hazard Symposium*, IBC, London, England (1987).
190. Stockton, G. W., S. J. Ehrlich-Moser, D. H. Chidester, and R. S. Wayne, "Versatile Macroscale Heat Flow Calorimeter for the Study of Chemical Processes," *Rev. Sci. Instrum.*, **57**, 3034–3042 (1986).
191. Fauske, H. K., "Emergency Relief System Design for Reactive and Non-Reactive Systems: Extension of the DIERS Methodology," *Plant/Operations Progress*, **7**, 153 (1988).
192. Proctor, P. and F. Nolan, "Easing the Pressure," *Process Engineering*, **71**, 35 (1990).
193. Rogers, R. L., "DEWAR Methods for Evaluating Chemical Reaction Hazards," DECHEMA Tagung, Bad Soden, Germany (1990).
194. Leung, J. C., H. K. Fauske, and H. G. Fisher, "Thermal Runaway Reactions in a Low Thermal Inertia Apparatus," *Thermochimica Acta*, **104**, 13 (1986).
195. Fauske, H. K., and J. C. Leung, "New Experimental Technique for Characterizing Runaway Chemical Reactions," *Chem. Eng. Prog.*, **81**, 39 (1985).
196. Technical Equipment Documentation RSST ("Quickly and Safely Evaluates Chemical Reaction and Assesses Process System Safety"), Fauske & Associates, Inc., Burr Ridge, IL.
197. Fauske, H. K., G. H. Clare, and M. J. Creed, "The Reactive System Screening Tool," ADPA International Joint Symposium, Virginia Beach, VA (1989).
198. Fauske, H. K., and M. J. Creed, "An Easy, Inexpensive Approach to DIERS Methodology," *Chem. Eng. Prog.*, **86**, No. 3, 45–49 (March 1990).

199. Trambouze, et al., *Chemical Reactors, Design/Engineering/Operation*, Editions Technip, Paris, France (1988).
200. Zwietering, T. N., "Suspending of Solid Particles in Liquid by Agitators," *Chem. Eng. Sci.*, **8**, 244 (1958).
201. Skelland, A. H. P., and R. Sektaria, "Minimum Impeller Speeds for Liquid–Liquid Dispersion in Baffled Vessels," *Ind. Eng. Chem. Process Des. Dev.*, **17**, No. 1 (1978).
202. Kurpiers, P., et al., "Zur Uberflutung Ein- und Zwei-stufiger Ruhrbehalter" ("About Flooding One- and Two-Stage Stirred Vessels"), *Chem. Eng. Tech.*, **57** (1985).
203. Jones, M. C., "Assessing a Runaway of a Grignard Reaction," *Plant/Operations Progress*, **8**, 200 (1989).
204. Bisio, A. and R. L. Kabel, *Scaleup of Chemical Processes*, John Wiley, New York, NY (1985).
205. Van Dierendonck, L. L., "Schaalvergroten, Ambacht of Wetenschap? Opschalen in de Bulkindustrie" ("Scale-up, Craft or Science? Scaling in Bulk Industry"), *Procestechnologie*, **2**, 20–28 (1990).
206. Kraume, M., and P. Zehner, "Suspendieren in Ruhrbehalter—Vergleich Unterschiedlicher Berechnungsgleichungen" ("Suspension of Solids in Stirred Tanks—Comparison of Different Calculation Methods"), *Chem. Ing. Tech.*, **60**, 822 (1988).
207. Choudhury, S., and L. Utiger, "Warmetransport in Ruhrkesseln: Scale-up Methoden" ("Heat Transfer in Stirred Tanks: Scale-up Methods"), *Chem. Ing. Tech.*, **62**, 154–155 (1990).
208. Steel, C. B. and P. F. Nolan, "Scale-up and Heat Transfer Data for Safe Reactor Operation" in *Proceedings of the International Symposium on Runaway Reactions*, p. 597, Center for Chemical Process Safety/AIChE, New York, NY (1989).
209. Thomas, P. H., "On the Thermal Conduction Equation for Self-Heating Materials with Surface Cooling," *Trans. Faraday Soc.*, **54**, 60 (1958).
210. Fisher, H. G. and D. D. Goetz, "Determination of Self-Accelerating Decomposition Temperatures for Self-Reactive Substances," *J. Loss Prev. Process Ind.*, **6**, No. 3, 183–194 (1993).
211. Fauske, H. K., "Scale-up for Safety Relief of Runaway Reactions," *Plant/Operations Progress*, **3**, 7 (1984).
212. Hendershot, D. C., "Safety Considerations in the Design of Batch Processing Plants," in *Proceedings of the International Symposium on the Prevention of Major Chemical Accidents*, page 3. 1, Center for Chemical Process Safety/AIChE, New York, NY (1987).
213. Singh, J., "Design of Process Plants Handling Highly Unstable and Flammable Materials by Use of Test Data," *Inst. Chem. Eng. Symposium Series*, **110**, 229 (1988).
214. Verhoeff, J., "Explosion Hazards of Tertiary Butyl Hydroperoxide (TBHP)," *Inst. Chem. Eng. Symposium Series*, **68** (1981).
215. Silverstein, J. L., B. D. Wood, and S. A. Leshaw, "Case Study in Reactor Design for Hazard Prevention," *J. Loss Prev.*, **78** (1981).
216. SAFIRE Computer Program (Tape) and Documentation, American Institute of Chemical Engneers, New York, NY (1993).

REFERENCES CITED 195

217. Tharmalingham, S., "The Evaluation of Self-Heating in Bulk Handling of Unstable Solids," in *Proceedings of the International Symposium on Runaway Reactions*, p. 293, Center for Chemical Process Safety/AIChE, New York, NY (1989).
218. Horner, R. A., "Direction of Plant Process Safety Regulations in the United States," *J. Loss Prev. Proc. Ind.*, **2** (1989).
219. Reay, D., "Fire and Explosion Hazards in Dryers," *Loss Prevention Bulletin*, **025**, 1.
220. Noronha, J. A., J. T. Merry, and W. C. Reid, "Deflagration Pressure Containment (DPC) for Vessel Safety Design," *Plant/Operations Progress*, **1**, 1 (1982).
221. Noronha, J. A., C. S. Brown, Jr., M. R. Juba, and J. Schmidt, "Kinetics of Runaway Reactions," CEP Technical Manual, *Loss Prevention*, **13** (1980), American Institute of Chemical Engineers, New York, NY.
222. Noronha, J. A., R. J. Seyler, and A. J. Torres, "Simplified Chemical and Equipment Screening for Emergency Venting Safety Reviews Based on the DIERS Technology," in *Proceedings of the International Symposium on Runaway Reactions*, Center for Chemical Process Safety/AIChE, New York, NY (1989).
223. Jercinovic, L. M., "A Laboratory Pressure Safety Program," *Plant/Operations Progress*, **3**, 34 (1984).
224. Welding, T. V., "Operational Experience with Total Containment Systems," *I. Chem. Eng. Symposium Series*, **85**, 251 (1984).
225. Hub, L., "On-Line Uberwachung von Exothermen Chemischen Prozessen mit dem OLIWA System" ("On-Line Control of Exothermic Processes by the OLIWA System"), *Chem. Eng. Tech.*, **54**, 181 (1982).
226. Hub, L., and J. D. Jones, "Early On-Line Detection of Exothermic Reactions," *Plant/Operations Progress*, **5**, 221 (1986).
227. Spence, J. P., and J. A. Noronha, "Reliable Detection of Runaway Reaction Precursion in Liquid Phase Reactions," *Plant/Operations Progress*, **7**, 234 (1985).
228. Wu, R. S. H., "Dynamic Thermal Analyzer for Monitoring Batch Processes," *Chem. Eng. Prog.*, **81**, No. 9, 57 (September 1985).
229. Helsby, G. B., and R. F. White, "Criteria for the Case in the Assessment and Control of Major Hazards," *I. Chem. Eng. Symposium Series*, **93** (1985).
230. Grossel, S. S., "An Overview of Equipment for Containment and Disposal of Emergency Relief System Effluents," *J. Loss Prev. Proc. Ind.*, **3**, No. 1, 112 (January 1990).
231. Kneal, M., "The Engineering of Relief Disposal—A Review Paper," *I. Chem. Eng. Symposium Series*, **85**, 183 (1984).
232. First, K. E., and J. E. Huff, "Design Charts for Two-Phase Flashing Flow in Emergency Pressure Relief Systems," *Plant/Operations Progress*, **8**, 40, (1989).
233. Huff, J. E., "The Role of Pressure Relief in Reactive Chemical Safety," in *Proceedings of the International Symposium on Prevention of Major Chemical Accidents*, p. 4. 43, Center for Chemical Process Safety/AIChE, New York, NY (1987).
234. Thomas, P. W., "Multi-Stage Over-Pressure Protection and Product Containment on High Pressure Polymerization Reactors," *I. Chem. Eng. Symposium Series*, **85**, 229 (1984).
235. Boeije, C. G., "Flare Relief Systems," *I. Chem. Eng. Symposium Series* (1979).

236. Perry, R. H. and D. W. Green (editor), "Perry's Chemical Engineers' Handbook," 6th Edition, McGraw-Hill, New York, NY (1984).
237. Speechly, et al., *I. Chem. Eng. Symposium Series*, (1979).
238. Prugh, R. H., and R. W. Johnson, *Guidelines for Vapor Release Mitigations*, Center for Chemical Process Safety/AIChE, New York, NY (1988).
239. Leung, J. C., A. C. Buckland, A. R. Jones, and L. D. Pesce, "Emergency Relief Requirements for Reactive Chemical Storage Tanks," *I. Chem. Eng. Symposium Series*, **110**, 169 (1988).
240. Fauske, H. K., M. A. Grolmes, and G. H. Clare, "Process Safety Evaluatioin Applying DIERS Methodology to Exisiting Plant Operations," *Plant/Operations Progress*, **8**, 19 (1989).
241. Grolmes, M. A. and J. C. Leung, "Code Method for Evaluating Integrated relief Phenomena," *Chem. Eng. Prog.*, **81**, 47 (1985).
242. Fauske, H. K., "Flashing Flows or Some Practical Guidelines for Emergency Releases," *Plant/Operations Progress*, **4**, 132 (1985).
243. Fauske, H. K., M. A. Grolmes, and J. C. Leung, "Multi-Phase Flow Considerations in Siting Emergency Relief Systems for Runaway Chemical Reactions," *Multi-Phase Flow and Heat Transfer III-B, Applications*, 899 (1984), Elsevier, Amsterdam, The Netherlands.
244. Boyle, W. J., "Sizing Relief Area for Polymerization Reactors," CEP Technical Manual, *Loss Prevention*, **1**, 78 (1967), American Institute of Chemical Engineers, New York, NY.
245. Leung, J. C., and H. K. Fauske, "Runaway System Characterization and Vent Sizing Based on DIERS Methodology," *Plant/Operations Progress*, **6**, 77 (1987).
246. Cox, R. A., "An Overview of Hazard Analysis," in *Proceedings of the International Symoisum on Prevention of Major Chemical Accidents*, p. 1. 37, Center for Chemical Process Safety/AIChE, New York, NY (1987).
247. Freeman, R. A., "The Use of Risk Assesment in the Chemical Industry," *Plant/Operations Progress*, **5**, 142 (1986).
248. Davies, K. R., "Techniques for the Identification and Assessment of Major Accident Hazards," *I. Chem. Eng. Symposium Series*, **93**, 289 (1985).
249. Lihou, D. A., "A Safe and Operable Plant at Minimum Cost," *I. Chem. Eng. Symposium Series*, **110** (1988).
250. Turney, R. D., "Designing Plants for the 1990s and Beyond," in *Proceedings of Safety and Loss Prevention in the Chemical & Oil Process Industries*, H2, Singapore (1989).
251. Lees, F. P., *Loss Prevention in the Process Industries*, Butterworth, London, England, and Boston, MA (1980).
252. *Dow's Fire and Explosion Index, Hazard Classification Guide*, Seventh Edition, Technical Manual, American Institute of Chemical Engineers, New York, NY (1994).
253. Lewis, D. J., "The Mond Fire, Explosion, and Toxicity Index Applied to Plant Layout and Spacing," CEP Technical Manual, *Loss Prevention*, **13**, 20 (1980), American Institute of Chemical Engineers, New York, NY.

254. Andreason, P., and B. Rasmussen, "Comparison of Methods of Hazard Identification at Plant Level," *J. Loss Prev. Proc. Ind.*, **3**, 339 (1990).
255. Cizek, J. G., "Diamond-Shamrock Loss Prevention Review Program," presented at Canadian Society for Chemical Engineering Conference (1982).
256. *Procedures for Performing a Failure Mode and Effect Analysis*, U. S. Department of the Navy, MIL-STD-1629A (1977), Washington, DC.
257. *A Guide to Hazard and Operability Studies*, Chemical Industries Association, Alembic House, London, England (1977).
258. Kletz, T. A., "HAZOP and HAZAN: Notes on the Identification and Assessment of Hazards," *I. Chem. Eng. Symposium Series* (1983).
259. Jones, M. C., and D. A. Lihou, "CAFOS—The Computer Aid for Operability Studies," *I. Chem. Eng. Symposium Series*, **97**, 249 (1986).
260. Soukas, J., "The Limitations of Safety and Risk Analysis," *I. Chem. Eng. Symposium Series*, **110**, 493 (1988).
261. *Risk Analysis in the Process Industries*, Institution of Chemical Engineers, Rugby, England, UK (1985).
262. Gibson, S. B., "Investment in Human Factors Pays Dividends," in *Proceedings of the International Symposium on Preventing Major Chemical Accidents*, p. 6. 41, Center for Chemical Process Safety/AIChE, New York, NY (1987).
263. Swain, A. D., "Relative Advantages of People and Machines in Process Industries," in *Proceedings of the International Symposium on Preventing Major Chemical Accidents*, p. 6. 97, Center for Chemical Process Safety/AIChE, New York, NY (1987).
264. Whalley, S. P., and J. K. Maund, "Improving Human Reliability in Design," *I. Chem. Eng. Symposium Series*, **97** (1986).
265. Fulton, A. S. and D. J. Barrett, "PES—An Opportunity for Better Safety Systems," *I. Chem. Eng. Symposium Series*, **97**, 215 (1986).
266. *An Investigation of Potential Hazards from Operations in the Canvey Island/Thurrock Area*, HMSO, London, England, UK (1978).
267. Roodbol, H. G., *Risk Analysis of Six Potentially Hazardous Objects in the Rijnmond Area—A Pilot Study*, D. Reidel Publishing Company, Dordrecht, The Netherlands, and Boston, MA (1982).
268. Pitblado, R. M., S. J. Shaw, and G. Stevens, "The SAFETY Risk Assessment Package and Case Study Applications," in *Proceedings of the Safety and Loss Prevention, Chemical and Oil Industries*, A4, Singapore (1989).
269. Boykin, R. F., and M. Kazarians, "Quantitative Risk Assessment for Chemical Operations," in *Proceedings of the International Symposium on Prevention of Major Chemical Accidents*, p. 1. 87, Center for Chemical Process Safety/AIChE, New York, NY (1987).
270. French, R. W., R. E. Olsen, and G. L. Peloquin, "Quantified Risk as a Decision Aid," in *Proceedings of the Safety and Loss Prevention, Chemical and Oil Industries*, G3, Singapore (1989)
271. Edmonson, J. N., "The Current State of Risk Assessment" in *Proceedings of the Safety and Loss Prevention, Chemical and Oil Industries*, G7, Singapore (1989).

272. "Process Safety Management of Highly Hazardous Chemicals, Explosives, and Blasting Caps," U. S. Occupational Safety and Health Administration, 57 *Federal Register* 6355, February 24, 1992.
273. European Council Directive on "Major Accidents of Certain Industrial Activities," ("Seveso Directive"), 82/501/EEC, June 4, 1982, as amended 87/216/EEC, March 19, 1987, European Union, Brussels, Belgium.
274. Responsible Care®, Chemical Manufacturers Association, Washington, DC (1988).
275. *Community Awareness and Emergency Response Handbook*, Chemical Manufacturers Association, Washington, DC (1989).
276. "Chemical Makers Pin Hopes on Responsible Care to Improve Image," *C&EN*, p. 13 (October 5, 1992).

SELECTED ADDITIONAL READINGS

Barton, J. A., and P. F. Nolan, "Runaway Reactions in Batch Reactors," *I. Chem. Eng. Symposium Series*, **85**, 13 (1984).

Beever, P. F., "Scaling Rules for Prediction of Thermal Ruanway," in *Proceedings of the International Symposium on Runaway Reactions*, p. 1, Center for Chemical Process Safety/AIChE, New York, NY (1989).

Capraro, M. A., and J. H. Strickland, "Preventing Fires and Explosions in Pilot Plants," *Plant/Operations Progress*, **8**, No. 4, 189 (April 1989).

Crowl, D. A., and J. F. Louvar, *Chemical Process Safety, Fundamentals with Applications*, Prentice-Hall, Englewood Cliffs, NJ (1990).

Crowl, D. A., and J. F. Louvar, "Instructional Videotapes on Chemical Process Safety," *Plant/Operations Progress*, **8**, 225 (1989).

Donnelly, R. E., "An Overview of OSHA's Process Management Standard (USA)," *Process Safety Progress*, **13**, No. 2, 53 (April 1994).

Gray, B. F., D. P. Coppersthwaite, and J. F. Griffiths, "A Novel, Thermal Instability in a Semi-Batch Reactor," *Process Safety Progress*, **12**, No. 1, 49 (January 1993).

Guidelines for Engineering Design for Process Safety, Center for Chemical Process Safety, American Institute of Chemical Engineers, New York, NY (1993).

Guidelines for Safe Automation of Chemical Processes, Center for Chemical Process Safety, American Institute of Chemical Engineers, New York, NY (1993).

Gustin, J. L., "Thermal Stability Screening and Reaction Calorimetry—Application to Runaway Hazard Assessment and Process Safety Management," *J. Loss Prev. Proc. Ind.*, **6**, 275 (1993).

Gygax, R, "Chemical Reaction Engineering for Safety," *Chem. Eng. Sci.*, **43**, No. 8, 1759–1771 (1988).

Heemskerk, A. H., "Detection and Avoidance of Explosion Hazards," DECHEMA Tagung, Bad Soden, Germany (1990).

Hugo, P. and J. Steinbach, "A Comparison of the Limits of Safe Operating of a SBR and CSTR," *Chem. Eng. Sci.*, **41**, No. 4, 1081–1087 (1986).

Kletz, T. A., *What Went Wrong? Case Histories of Process Plant Disasters*, Gulf Publishing Company, Houston, TX (1985).

Kletz, T. A., *Learning from Accidents in Industry*, Butterworth, London, England, and Boston, MA (1988).

LaVine, R., *Guidelines for Safe Storage and Handling of High Toxic Hazard Materials*, Center for Chemical Process Safety, American Institute of Chemical Engineers, New York, NY (1988).

L. B. Levy, "The Effect of Oxygen on Vinyl Acetate and Acrylic Monomer Stabilization," *Process Safety Progress*, **12**, No. 1, 47 (January 1993).

Munson, R. E., "Process Hazards Management in DuPont," *Plant/Operations Progress*, **4** (1985).

Paul, E. L., "Design of Reaction Systems for Specialty Organic Chemicals," *Chem. Eng. Sci.*, **43**, No. 8, 1773–1782 (1988).

Steensma, M., and K. R. Westerterp, "Thermally Safe Operation of a Semi-Batch Reactor for Liquid-Liquid Reactions—Slow Reactions," *Ind. Eng. Chem. Res.*, **29**, 1259–1270 (1990).

Uehara, Y., and A. Kitamura, "Evaluation of Potential Hazards of Oxidizing Substances by Burning Tests," in *Proceedings of the International Conference on Safety and Loss Prevention in Chemical and Oil Processing Industries*, D5, Singapore (1989).

Wade, D. E., "Reduction of Risks by Reduction of Toxic Material Inventories," in *Proceedings of the International Symposium on Prevention of Major Chemical Accidents*, p. 2. 1, Center for Chemical Process Safety/AIChE, New York, NY (1987).

Wang, S. S. Y., S. Kiang, and W. Merkl, "Investigation of a Thermal Runaway Hazard—Drum Storage of Thionyl Chloride/Ethyl Acetate Mixture," *Process Safety Progress*, **13**, No. 3, 153 (July 1994).

Wilday, A., "The Safe Design of Chemical Plants with No Need for Pressure Relief," *I. Chem. Eng. Symposium Series*, **124**, 243–253 (1991).

INDEX

A

Accelerating rate calorimeter (ARC)
  described, 71–76
  reactive system process safety results, 137
  scale-up and pilot plants results, 145
  test data examples, 24–25
  thermal stability testing, 18, 19
Accountability, management program elements, 180
Acetic acid, oxygen balance of, 34
Acetone, vapor pressure of, 108
Acetonitrile, enthalpy of formation, 37
Acetylene, structure of, 30, 32
Aci-nitro salts, structure of, 32
Acyl nitrates, structure of, 32
Acyl nitrites, structure of, 32
Adiabatic calorimetry, typical curves from, 23
Adiabatic storage test (AST)
  described, 66–71
  powder stability tests, 76
  thermal stability testing, 18, 19, 20
Adiabatic temperature increase, energy content management, 101–102
Aerated powder test, powder stability tests, 77
Agitation, energy content management, 106–107
Air (hazard identification)
  instability/incompatibility factors, 49
  redox systems, 50
Alkyl hydroperoxides, structure of, 32

Alkyl nitrates, structure of, 32
Alkyl nitrites, structure of, 32
Allene, structure of, 30
Amine chromium peroxocomplexes, structure of, 32
Amine metal oxo salts, structure of, 32
Arenediazo aryl sulfides, structure of, 32
Arenediazoates, structure of, 32
Assessment strategies, chemical reactivity, 6–8
Audits, management program elements, 181
Autocatalytic decomposition, thermal stability test examples, 25–26
Autoclave tests
  described, 80–83
  thermal stability testing, 19
Average bond energy summation method, hazard identification, enthalpy calculation, 35–36
Azides, structure of, 32
Azo groups, high-energy substance identification, 31, 32
N-Azolium nitroimidates, structure of, 32

B

Batch reactor
  energy content management, 108–113
  process design applications, 148–149
Batch/tank reactor, bench-scale equipment for, data acquisition and use, 116–129. *See also* Bench-scale equipment

201

Batch-to-continous process, process design applications example, 154
Bench-scale equipment (for batch/tank reactors), 116–129
  Contalab, 119–121
  overview of, 116–117
  quantitative reaction calorimeter, 122–123
  Reaction Calorimeter (RC1), 117–119
  Reactive System Screening Tool (RSST), 126–129
  specialized reactors, 123–124
  ThermoMetric instruments, 121–122
  Vent Size Package (VSP), 124–126
Benzene, decomposition product, 38
Benzotriazole, structure of, 30
Biphenyl, decomposition product, 38
Bis-arenediazo oxides, structure of, 32
Bis-arenediazo sulfides, structure of, 32
Boiling Liquid Expanding Vapor Explosion (BLEVE), 11, 156–157
Bowes and Cameron test, powder stability tests, 76
1,3-Butadiene
  enthalpy of formation, 37
  structure of, 30
Butanone, decomposition product, 38
t-Butyl hydroperoxide, enthalpy of formation, 37
t-Butyl methyl sulfide, enthalpy of formation, 37
t-Butylperoxybenzoate
  decomposition products of, 38–39
  deflagration testing, 81–82

C

Capital project review, management program elements, 180
Carbon dioxide, decomposition product of, 38
Catalytic effects, hazard identification, instability/incompatibility factors, 48
Change management management program elements, 180
Chemical energy density, CHETAH program, 41

Chemical kinetics, scale-up and pilot plants, 139–140
Chemical reactivity
  assessment and testing strategies, 6–8
  classes of materials, 1–2
  defined, 2
  detonations, deflagrations, and runaway reactions, 5–6
  generally, 4–5
  hazardous, identification of, 9–88. *See also* Hazard identification
  process/reactor design and, 2–3, 89–173. *See also* Process/reactor design
CHETAH program, 116
  compared to other programs, 41
  containment and, 161–162
  described, 40–44
  hazard identification
    enthalpy calculation, 36, 37, 38, 39
    test strategies, 16
Chronology, for testing, 7–8
Codes, management program elements, 181
Combustibility testing, 14
Concentration
  energy content management, 104
  hazard identification, instability/incompatibility factors, 48
Confinement
  hazard identification, instability/incompatibility factors, 49
  sensitivity testing for, described, 86
Conservation of energy law, energy content management, 100–108
Constant heating rate tests, typical curves from, 23
Containment, 159–164
  full scale example, 164
  gas-vapor release determination, 160–161
  generally, 159–160
  laboratory scale example, 161–164
  mitigation measures, 172–173
Contalab, described, 119–121
Contamination, reactive system process safety, 135–136

INDEX

Continuous nitrogen, process design applications example, 151–153
Continuous reactor systems, process/reactor design, energy content management, 108–113
Continuous stirred tank reactor, 109–113, 139
Cooling loss, reactive system process safety, 135
Corrective action, management program elements, 181
CPA ThermoMetric instruments, described, 121–122
Cyanogen, structure of, 30
Cyclopropane
  enthalpy of formation, 37
  hazard identification, enthalpy calculation, 36

D

Data acquisition and use, 116–159
  bench-scale equipment for batch/tank reactors, 116–129
    Contalab, 119–121
    CPA ThermoMetric instruments, 121–122
    generally, 116–117
    quantitative reaction calorimeter, 122–123
    Reaction Calorimeter (RC1), 117–119
    Reactive System Screening Tool (RSST), 126–129
    specialized reactors, 123–124
    Vent Size Package (VSP), 124–126
  dryers and filters, 157–159
  generally, 116
  process design applications, 147–154
    batch and semi-batch plants, 148–149
    batch-to-continous example, 154
    continuous nitrogen example, 151–153
    generally, 147
    integrated relief evaluation, 154
    peroxides example, 149–151
    self-heating example, 153–154
  reactive system process safety, 129–137
    malfunction and process deviation testing, 134–136
    physical states of system, 131–132
    pressure effect, 137
    results from ARC, RSST, and VSP, 137
    test plan, 129–131
    test results, 132–134
  scale-up and pilot plants, 137–147
    accelerating rate calorimeter (ARC) results, 145
    chemical kinetics, 139–140
    generally, 137–139
    heat transfer, 141–142
    mass transfer/mixing, 140–141
    self-heating, 142–145
    Vent Size Package (VSP) results, 145–146
  storage and handling, 154–157
    passive prevention methods, 156–157
    peroxides example, 155–156
    scale-up example, 154–155
Deflagration
  described, 5–6
  exothermic reactions, 11–12
  hazard identification test strategies, 15–16
  testing for, described, 80–83
Detection, runaways, protective measures, 164–167
Detonation
  described, 5–6
  exothermic reactions, 11–12
  hazard identification test strategies, 15–16
  testing for, described, 78–80
Dewar flask test
  described, 66–71
  powder stability tests, 76
  thermal stability testing, 18, 19, 20
Diazeno compounds, structure of, 32
Diazo compounds, structure of, 32
Diazomethane, structure of, 30
Diazonium carboxylates and salts, structure of, 32
Diazonium sulfides, structure of, 32

Dichloroacetic acid, enthalpy of formation, 37
Differential scanning calorimetry (DSC)
  described, 52–59
  screening tests, 13–14
  test data examples, 24, 25, 26
Differential thermal analysis (DTA)
  described, 52–59
  screening tests, 13–14
  typical curves from, 23
Difluoro amino compounds, structure of, 32
Dinitromethane, oxygen balance of, 34
Dispersion, mitigation measures, 172–173
Documentation, management program elements, 180
Dow and Mond Hazard Indexes, hazard evaluation procedures, 176
Dryers, data acquisition and use, 157–159
Dutch pressure vessel test, confinement sensitivity, 86

E
Education, management program elements, 181
Energetic substances, potentially explosive criteria, 14–15
Energy, process design and, 2
Enthalpy
  calculation of, hazard identification, thermodynamic calculations, 35–39
  runaway reactions and, 93–94
  testing strategies, 15
  thermal testing, 57–58
1,2-Epoxides, structure of, 32
Equipment design, process design and, 3
Equipment integrity, management program elements, 181
Ethane benzene, decomposition product, 38
Ethylenediamine, enthalpy of formation, 37
Exothermic reactions
  hazard identification, 11–13
  thermodynamics and, 28–30

Explosibility testing, 78–86
  confinement sensitivity, 86
  deflagration testing and autoclave testing, 80–83
  detonation testing, 78–80
  mechanical sensitivity testing, 83–85
Explosion, exothermic reactions, 11–13

F
Failure logic diagram, hazard evaluation procedures, 177–178
Failure modes, effects, and criticality analysis (FMECA), hazard evaluation procedures, 177
Filters, data acquisition and use, 157–159
Flammability testing
  described, 88
  hazard identification test strategies, 19
Flaring, mitigation measures, 172–173
Flash point testing, 14
Fluoro dinitromethyl compounds, structure of, 32
Friction sensitivity test, mechanical sensitivity testing, 83
Full scale production, assessment and testing strategies, 7

G
Gas-vapor release determination, containment, 160–161
Gibbs free energy
  thermodynamics and, 28–30
  TIGER program, 46

H
Haloacetylene derivatives, structure of, 32
Halo-aryl metal compounds, structure of, 32
Halogenated hydrocarbons/metal reactions, hazard identification, instability/incompatibility factors, 52
Halogen azides, structure of, 32
Hazard evaluation procedures, management, 176–179
Hazard identification, 9–88
  exothermic reactions, 11–13

experimental thermal and reactivity measurements, 13
high-energy substance identification, 30–33
instability/incompatibility factors, 46–52
  generally, 46–49
  halogenated hydrocarbons/metal reactions, 52
  redox systems, 49–51
  water reactions, 51–52
management, 175–176
strategy for, 9–11
testing, 13–20, 52–88. See also Testing
  detonation and deflagration criteria, 15–16
  potentially explosive criteria, 14–15
  reactive substances, 19–20
  screening tests, 13–14
  thermal stability, 16–19
thermodynamic calculations, 33–46
  computer program application, 39–46
  enthalpy calculation, 35–39
  oxygen balance, 33–35
  thermodynamics and, 28–30
Hazard Operability Study (HAZOP)
  hazard evaluation procedures, 177
  process/reactor design, 99
Heat
  exothermic reactions, 11–13
  hazard identification, 10
Heat effect, chemical reactivity, 5
Heat flux differential scanning calorimetry (DSC), 52–53
Heat transfer
  process/reactor design, energy content management, 113–115
  scale-up and pilot plants, 141–142
High-energy substance identification, hazard identification, 30–33
Hot plate tests, powder stability tests, 77
Human factors
  error analysis, hazard evaluation procedures, 178
  management program elements, 181
Hydrogen cyanide, structure of, 30

Hydroxyammonium salts, structure of, 32

I
Impact testing, mechanical sensitivity testing, 83–84
Impurities, hazard identification, instability/incompatibility factors, 48
Incident investigation, management program elements, 181
Inhibitor depletion, thermal stability test examples, 25–26
Instability/incompatibility factors, hazard identification, 46–52
  generally, 46–49
  halogenated hydrocarbons/metal reactions, 52
  redox systems, 49–51
  water reactions, 51–52
Instrumentation, runaways, protective measures, 164–167
Integrated relief evaluation, process design applications, 154
Isoperibolic calorimetry
  described, 59–61
  thermal stability testing, equipment, 18
Isothermal storage test (IST)
  described, 62–66
  thermal stability testing, 18
  typical curves from, 23

K
Kinetics, thermal testing, 58
Koenen test, confinement sensitivity, 86

L
Laboratory scale example, containment, 161–164
Law, management program elements, 181
Law of conservation of energy, energy content management, 100–108
Layer test, powder stability tests, 77

M
Malfunction testing, reactive system process safety, 134–136

Management, 175–182
  future trends, 181–182
  hazard evaluation procedures, 176–179
  hazard identification and quantification, 175–176
  program elements, 180–181
Mass balance, energy content management, 107–108
Mass transfer/mixing, scale-up and pilot plants, 140–141
Mechanical sensitivity testing, described, 83–85
Mechanical shock test, mechanical sensitivity testing, 83–84
Metal acetylides, structure of, 32
N-metal derivatives, structure of, 32
Metal fulminates, structure of, 32
Metal/halogenated hydrocarbon reactions, hazard identification, instability/incompatibility factors, 52
Metal peroxides, structure of, 32
Methane, decomposition product, 38
1-Methyl-2-propanol, decomposition product, 38
Mischarging, reactive system process safety, 136
Mitigation measures, 168–173
  dispersion, flaring, scrubbing, and containment, 172–173
  reaction quenching methods, 168–169
  relief disposal, 170–172
  sulfonation example, 169–170
  venting, 173

N

NASA-CET program
  compared to other programs, 41
  described, 46
Nitroalkanes, structure of, 32
N-nitro compounds, structure of, 32
Nitroethane, enthalpy of formation, 37
Nitrogen trichloride, structure of, 30
Nitro groups, high-energy substance identification, 31
Nitroso compounds, structure of, 32
N-nitroso compounds, structure of, 32

Noise suppression, methods of, 167
NOTS-CRUISE, 39

O

On-Line Warning System (OLIWA), 164
Onset temperature, thermal stability test examples, 22–25, 26
Organic compounds, hazard identification, redox systems, 51
Oxidizers, hazard identification, redox systems, 51
Oxidizing properties, reactivity testing, 87–88
Oxygen balance
  CHETAH program, 42, 44
  hazard identification, 10, 33–35

P

Passive prevention methods, storage and handling, 156–157
Performance, management program elements, 181
Perioxidation, hazard identification, redox systems, 50
Peroxides
  high-energy substance identification, 31, 33
  process design applications example, 149–151
  storage and handling example, 155–156
  structure of, 32
Peroxoacid salts, structure of, 32
Peroxyacids, structure of, 32
Peroxysters, structure of, 32
Phenyl, decomposition product, 38
1-Phenyl-2-propanone, decomposition product, 38
Pilot plant, assessment and testing strategies, 6. *See also* Scale-up and pilot plants
Plug flow reactor, 109–113
Polynitro alkyl compounds, structure of, 32
Polynitro aryl compounds, structure of, 32
Potentially explosive criteria, hazard identification test strategies, 14–15

Powder bulk test, powder stability tests, 77
Powders, hazard identification test strategies, 19–20
Powder stability tests, described, 76–78
Power-compensation differential scanning calorimetry (DSC), 53
Preliminary hazard analysis (PHA), hazard evaluation procedures, 177
Pressure effect, reactive system process safety, 137
Probability correlation, CHETAH program, 41–42
Process design applications
  batch and semi-batch plants, 148–149
  batch-to-continous example, 154
  continuous nitrogen example, 151–153
  generally, 147
  integrated relief evaluation, 154
  peroxides example, 149–151
  self-heating example, 153–154
Process deviation testing, reactive system process safety, 134–136
Process hazard evaluation, 98
Process knowledge, management program elements, 180
Process/reactor design, 89–173
  data acquisition and use, 116–159. *See also* Data acquisition and use
  energy content management, 100–115
  heat transfer, 113–115
  law of conservation of energy, 100–108
  reactor choice, 108–113
  equipment requirements, 100
  overview of, 89–90
  protective measures, 159–173
  containment, 159–164
  mitigation measures, 168–173
  runaways, instrumentation and detection of, 164–167
  safety strategy and rules, 96–99
  thermal hazards, 90–100
Process risk management, management program elements, 180
Process system checklist, hazard evaluation procedures, 176
Propanone, decomposition product, 38

Protective measures
  containment, 159–164
  full scale example, 164
  gas-vapor release determination, 160–161
  generally, 159–160
  laboratory scale example, 161–164
  mitigation measures, 168–173
  dispersion, flaring, scrubbing, and containment, 172–173
  reaction quenching methods, 168–169
  relief disposal, 170–172
  sulfonation example, 169–170
  venting, 173
  runaways, instrumentation and detection of, 164–167
Pryophoric properties, reactivity testing, 87
Pyrimidine, enthalpy of formation, 37, 38

Q
Quantitative reaction calorimeter, described, 122–123
Quantitative risk assessment procedure, hazard evaluation procedures, 179

R
Reaction Calorimeter (RC1), described, 117–119
Reaction quenching methods, mitigation measures, 168–169
Reaction rate
  energy content management, 102–104
  process design and, 2–3
Reactive chemical. *See* Chemical reactivity
Reactive substances, hazard identification test strategies, 19–20
Reactive system process safety
  malfunction and process deviation testing, 134–136
  physical states of system, 131–132
  pressure effect, 137
  results from ARC, RSST, and VSP, 137
  test plan, 129–131
  test results, 132–134

Reactive system screening tool (RSST)
  described, 126–129
  reactive system process safety results, 137
  thermal stability testing, 18, 19
Reactivity testing, 87–88
  oxidizing properties, 87–88
  pryophoric properties, 87
  water, 87
Reactor choice, process/reactor design, energy content management, 108–113
Reactor design. See Process/reactor design
Record keeping, management program elements, 180
Redox systems, hazard identification, instability/incompatibility factors, 49–51
REITP2
  compared to other programs, 41
  described, 44–45
Relief disposal, mitigation measures, 170–172
Ring stress, hazard identification, enthalpy calculation, 36
Runaway reactions. See also Stability/runaway hazard assessment report
  causes of, 90–92
  described, 5–6
  exothermic reactions, 11–12
  prevention of, 1, 92–100
  protective measures, instrumentation and detection of, 164–167
  testing for. See Thermal stability and runaway testing

S

Scale-up and pilot plants
  accelerating rate calorimeter (ARC) results, 145
  assessment and testing strategies, 6
  chemical kinetics, 139–140
  generally, 137–139
  heat transfer, 141–142
  mass transfer/mixing, 140–141
  self-heating, 142–145

storage and handling example, 154–155
Vent Size Package (VSP) results, 145–146
Screening tests, 52–61. See also Testing
  differential thermal analysis/differential scanning calorimetry, 52–59
  hazard identification test strategies, 13–14
  isoperibolic calorimetry, 59–61
Scrubbing, mitigation measures, 172–173
Self accelerating decomposition temperature (SADT) values
  Dewar flask testing, 67
  high-energy substance identification, 31, 33
Self-heating
  process design applications example, 153–154
  scale-up and pilot plants, 142–145
Semi-batch plants, 109–113, 148–149
Shockwave propagation
  detonation testing, 78–79
  hazard identification test strategies, 15–16
Solvents, hazard identification, instability/incompatibility factors, 49
Specialized reactors, described, 123–124
Stability. See Thermal stability
Stability/runaway hazard assessment report, thermal stability test examples, 26–28
Standards, management program elements, 181
Steel sleeve test, confinement sensitivity, 86
Steric hindrance, hazard identification, enthalpy calculation, 36
Storage and handling
  data acquisition and use, 154–157
  hazard identification test strategies, 19
  passive prevention methods, 156–157
  peroxides example, 155–156
  scale-up example, 154–155
  thermal stability, hazard identification test strategies, 16–17
Substance identification. See High-energy substance identification

Sulfonation example, mitigation measures, 169–170
Surface fouling, energy content management, 106–107

T

Temperature, hazard identification, instability/incompatibility factors, 46, 48
Testing, 52–88. *See also* Screening tests
  explosibility testing, 78–86
    confinement sensitivity, 86
    deflagration testing and autoclave testing, 80–83
    detonation testing, 78–80
    mechanical sensitivity testing, 83–85
  flammability testing, 88
  reactivity testing, 87–88
    oxidizing properties, 87–88
    pryophoric properties, 87
    water, 87
  screening tests, 52–61
    differential thermal analysis/ differential scanning calorimetry, 52–59
    isoperibolic calorimetry, 59–61
  strategies for, chemical reactivity, 6–8
  thermal stability and runaway testing, 61–78
    accelerating rate calorimeter (ARC), 71–76
    Dewar flask and adiabatic storage tests, 66–71
    generally, 61–62
    isothermal storage tests, 62–66
    powder stability tests, 76–78
Test plan, reactive system process safety, 129–131
Tetrazoles, structure of, 32
TGAP, 39
THEDIC, 39
Thermal activity monitor (TAM), thermal stability testing, 18
Thermal analysis. *See* Differential scanning calorimetry (DSC); Differential thermal analysis (DTA)
Thermal explosion, exothermic reactions, 12–13
Thermal hazards, process/reactor design, 90–100
Thermal runaway. *See* Runaway reactions
Thermal stability, hazard identification test strategies, 16–19
Thermal stability and runaway testing, 61–78
  accelerating rate calorimeter (ARC), 71–76
  Dewar flask and adiabatic storage tests, 66–71
  generally, 61–62
  isothermal storage tests, 62–66
  powder stability tests, 76–78
Thermal stability screening test. *See* Screening tests
Thermal stability test methods, 20–28
  examples of, 22–28
  overview of, 20–21
Thermodynamic calculations, hazard identification, 33–46
  computer program application, 39–46
  enthalpy calculation, 35–39
  oxygen balance, 33–35
Thermodynamics, hazard identification and, 28–30
ThermoMetric instruments, described, 121–122
TIGER program
  compared to other programs, 41
  described, 45–46
Time-to-runaway, adiabatic temperature increase, 102
TNT, hazard identification, enthalpy calculation, 36
Toluene, decomposition product, 38
Training, management program elements, 181
Triazines, structure of, 32
N,N,N-trifluoroalkylimidines, structure of, 32
2,4,6-Trinitrotoluene, enthalpy of formation, 37

U

United Nations DDT tests, 82
United Nations deflagration test, 80

United Nations flammability testing, 88
United Nations powder stability tests, 76–77
United Nations trough test, 80

V
Velocity of propagation. *See* Shockwave propagation
Venting, mitigation measures, 173
Vent Size Package (VSP)
  described, 124–126
  reactive system process safety results, 137
  scale-up and pilot plants results, 145–146

W
Water
  hazard identification, instability/incompatibility factors, 51–52
  reactivity testing, 87
What-if analysis, hazard evaluation procedures, 177

Y
"Y" criterion, CHETAH program, 43–44